Ontario Geological Survey
Report 223

Geology of the Manitou Lakes Area District of Kenora

(Stratigraphy and Petrochemistry)

by
C.E. Blackburn

1982

Ontario
Ministry of Natural Resources
Hon. Alan W. Pope
Minister
W. T. Foster
Deputy Minister

Printed in Canada

ISSN 0704-2582
ISBN 0-7743-7063-7

Publications of the Ontario Ministry of Natural Resources are available from the following sources. Orders for publications should be accompanied by cheque or money order payable to the *Treasurer of Ontario*.

Reports, maps, and price lists (personal shopping or mail order):

Public Service Centre, Ministry of Natural Resources
Room 1640, Whitney Block, Queen's Park
Toronto, Ontario M7A 1W3

Reports only (personal shopping):

Ontario Government Bookstore
Main Floor, 880 Bay Street
Toronto, Ontario

Reports and maps (mail order or telephone orders):

Publications Services Section, Ministry of Government Services
5th Floor, 880 Bay Street
Toronto, Ontario M7A 1N8
Telephone (local calls), 965-6015
Toll-free long distance, 1-800-268-7540
Toll-free from area code 807, 0-ZENITH-67200

Blackburn, C.E.
1982: Geology of the Manitou Lakes Area, District of Kenora, (Stratigraphy and Petrochemistry); Ontario Geological Survey Report 223, 61p. Accompanied by Map 2476, scale 1:50,000.

1000-300-82-CAR

CONTENTS

Abstract vii
Introduction 1
Present Geological Survey and Compilation 2
Access 2
Previous and Current Geological Work 2
Acknowledgments 3
General Geology 3
Table of Lithologic Units 4
Early Precambrian 5
Metavolcanics 5
Subalkalic Mafic Metavolcanics 6
Subalkalic Intermediate Metavolcanics 7
Subalkalic Felsic Metavolcanics 7
Alkalic Mafic Metavolcanics 8
Metasediments 8
Clastic Metasediments 9
Chemical Metasediments 10
Metamorphosed Mafic and Ultramafic Intrusive Rocks 10
Felsic and Intermediate Intrusive Rocks 11
Hypabyssal Intrusions 11
Batholithic Granitic Complexes 11
Intrabelt Plutons 14
Middle to Late Precambrian 15
Diabase Dikes 15
Stratigraphy of the Manitou-Stormy Lakes Volcano-Sedimentary Belt 15
Southeast of the Manitou Straits Fault 16
Wapageisi Lake Group 16
Starshine Lake Sub-Group 18
Etta Lake Formation 19
Glass Bay Formation 19
Manitou Group 19
Cane Lake Formation 20
Sunshine Lake Subvolcanic Intrusion 20
Sunshine Lake Formation and Associated Sills 21
Uphill Lake Formation 21
Mosher Bay Formation 21
Stormy Lake Group 22
Thundercloud Lake Subvolcanic Intrusion 22
Boyer Lake Group 23
Gabbro Lake, Mountdew Lake, Mosher Bay and Kenny Lake Gabbros 23
Northwest of the Manitou Straits Fault 23
Blanchard Lake Group 23
Upper Manitou Lake Group 24
Frenchman Island Subvolcanic Intrusion 24
Pincher Lake Group 24
Stony Island Formation 25
Garnet Bay Formation 25
Dog Lake Formation 25
Noonan Lake Formation 26
Johnar Lake Formation 26
Manitumeig Lake Formation 26
Benson Bay Sub-Group 26
Beaverhead Island Formation 26
Mitchell Lake Gabbro 27

Petrochemistry 27
Volcano-Sedimentary Belt 27
Wapageisi Lake Group 27
Manitou Group 35
Stormy Lake Group 40
Boyer Lake Group 41
Blanchard Lake Group 41
Upper Manitou Lake Group 42
Pincher Lake Group 42
Granitic Rocks 42
Batholithic Granitic Complexes 42
Meggisi Pluton 42
Intrabelt Plutons 44
Taylor Lake Stock 44
Scattergood Lake and Carleton Lake Stocks 45
Metamorphism 45
Structural Geology 45
Major Faults 45
Manitou Straits Fault 46
Mosher Bay-Washeibemaga Lake Fault 46
Major Folds 47
Manitou Anticline 47
Kamanatogama Syncline 48
Geologic Evolution of the Manitou Lakes Area 49
Economic Geology 50
Gold Deposits 50
Base Metal Sulphide Deposits 53
Molybdenum Deposits 55
Iron Deposits 55
References 56
Index 59

TABLES

1—Table of lithologic units 4
2—Comparative table for mineralogy and tectonic setting of granitoid rocks 12
3—Correlation of stratigraphic units with lithologic units of previous reports 17
4—Chemical analyses of metavolcanics 28
5—Chemical analyses of samples from the Sunshine Lake Formation and associated sills 39
6—Average chemical compositions of granitic and porphyry bodies 43
7—Association and genesis of gold deposits 52
8—Exploration, mineralization and association of base metal sulphide deposits 54

FIGURES

1—Key map showing location of the Manitou Lakes area. vii
2—Simplified sandstone classification. 9
3—Stratigraphic map of the Manitou Lakes area. Chart A
4—Cation and AFM plots for metavolcanics of Wapageisi Lake, Boyer Lake, Upper Manitou Lake, and Pincher Lake groups. 30
5—Cation and AFM plots for Beck, Starshine, Doyle, and Jonas Lakes sections. 32
6—Variation of composition with stratigraphic height for Beck, Starshine, Doyle, and Jonas Lakes sections. 34
7—Chemical classification of samples from the Sunshine Lake Formation. 38
8—Chemical variation in subvolcanic felsic porphyries and associated flows, and intermediate pyroclastics of the Stormy Lake group. 40
9—Chemical variation in granitic rocks and associated felsic porphyries. 44
10—Structural geometry of the Manitou Anticline: primary structures and cross sections. Chart B
11—Structural geometry of the Manitou Anticline: foliation trends and dip contours. Chart B
12—Structural geometry of the Manitou Anticline: lineation trends and plunge contours. Chart B

MAP

(back pocket)
Map 2476 (coloured) — Upper Manitou Lakes — Sunshine Lake, District of Kenora. Scale 1:50 000.

CHARTS

Chart A — Figure 3.
Chart B — Figures 10,11,12.

Conversion Factors for Measurements in Ontario Geological Survey Publications

If the reader wishes to convert imperial units to SI (metric) units or SI units to imperial units the following multipliers should be used:

CONVERSION FROM SI TO IMPERIAL			CONVERSION FROM IMPERIAL TO SI		
SI Unit	*Multiplied by*	*Gives*	*Imperial Unit*	*Multiplied by*	*Gives*
			LENGTH		
1 mm	0.039 37	inches	1 inch	**25.4**	mm
1 cm	0.393 70	inches	1 inch	**2.54**	cm
1 m	3.280 84	feet	1 foot	**0.304 8**	m
1 m	0.049 709 7	chains	1 chain	20.116 8	m
1 km	0.621 371	miles (statute)	1 mile (statute)	**1.609 344**	km
			AREA		
1 cm^2	0.155 0	square inches	1 square inch	**6.451 6**	cm^2
1 m^2	10.763 9	square feet	1 square foot	**0.092 903 04**	m^2
1 km^2	0.386 10	square miles	1 square mile	2.589 988	km^2
1 ha	2.471 054	acres	1 acre	0.404 685 6	ha
			VOLUME		
1 cm^3	0.061 02	cubic inches	1 cubic inch	**16.387 064**	cm^3
1 m^3	35.314 7	cubic feet	1 cubic foot	0.028 316 85	m^3
1 m^3	1.308 0	cubic yards	1 cubic yard	0.764 555	m^3
			CAPACITY		
1 L	1.759 755	pints	1 pint	0.568 261	L
1 L	0.879 877	quarts	1 quart	1.136 522	L
1 L	0.219 969	gallons	1 gallon	**4.546 090**	L
			MASS		
1 g	0.035 273 96	ounces (avdp)	1 ounce (avdp)	28.349 523	g
1 g	0.032 150 75	ounces (troy)	1 ounce (troy)	**31.103 476 8**	g
1 kg	2.204 62	pounds (avdp)	1 pound (avdp)	**0.453 592 37**	kg
1 kg	0.001 102 3	tons (short)	1 ton (short)	**907.184 74**	kg
1 t	1.102 311	tons (short)	1 ton (short)	**0.907 184 74**	t
1 kg	0.000 984 21	tons (long)	1 ton (long)	**1016.046 908 8**	kg
1 t	0.984 206 5	tons (long)	1 ton (long)	**1.016 046 908 8**	t
			CONCENTRATION		
1 g/t	0.029 166 6	ounce (troy)/ ton (short)	1 ounce (troy)/ ton (short)	34.285 714 2	g/t
1 g/t	0.583 333 33	pennyweights/ ton (short)	1 pennyweight/ ton (short)	1.714 285 7	g/t

OTHER USEFUL CONVERSION FACTORS

1 ounce (troy)/ton (short)	20.0	pennyweights/ton (short)
1 pennyweight/ton (short)	0.05	ounce (troy)/ton (short)

NOTE—Conversion factors which are in bold type are exact. The conversion factors have been taken from or have been derived from factors given in the Metric Practice Guide for the Canadian Mining and Metallurgical Industries published by The Mining Association of Canada in cooperation with the Coal Association of Canada.

ABSTRACT

This report is a synopsis of the geology of the Manitou Lakes area, located about 45 km south of the town of Dryden, and comprising an area of about 1016 km^2. Emphasis is placed on stratigraphy, petrochemistry, structure and mineralization of the volcano-sedimentary belt, and petrochemistry of certain granitic rocks.

Early Precambrian metavolcanics, metasediments and subvolcanic intrusive rocks are assigned to seven groups. Four of these groups are southeast of the Manitou Straits Fault, and three are northwest of it. In ascending structural order, the former are as follows. The Wapageisi Lake group comprises a thick, lower subgroup of tholeiitic basalt with komatiitic trend and upward passage from magnesian to Fe-rich, an overlying mixed pyroclastic-epiclastic-mafic volcanic formation, and an upper mafic volcanic formation. The Manitou group comprises a lower calc-alkaline, dacitic to andesitic pyroclastic formation intruded by a subvolcanic calc-alkaline dacitic porphyry body, an overlying marginally alkalic trachybasaltic volcanic formation and associated sills, and two upper epiclastic formations. The Stormy Lake group, a lateral equivalent of the Manitou group, is a heterogeneous assemblage of calc-alkaline andesitic to dacitic pyroclastic and epiclastic rocks, and has been intruded by a subvolcanic calc-alkaline dacitic to rhyolitic porphyry body. The Boyer Lake group comprises predominantly tholeiitic Fe-rich basalt flows, with minor calc-alkaline dacite, intruded by mafic to ultramafic sills and sill-like bodies. The Wapageisi Lake group is overlain conformably by the Manitou and Stormy Lake groups; the Boyer Lake group structurally overlies the Manitou and Stormy Lake groups along the Mosher Bay – Washeibemaga Lake Fault.

The three groups northwest of the major fault are as follows, in ascending structural and conformable stratigraphic order. The Blanchard Lake group comprises predominantly tholeiitic Fe-rich basalt flows. The Upper Manitou Lake group is a heterogeneous assemblage of calc-alkaline andesitic to dacitic pyroclastics, and mafic flows, intruded by a subvolcanic calc-alkaline granodioritic to quartz dioritic porphyry body. The Pincher Lake group comprises seven formations which are composed alternately of mafic flows and intermediate pyroclastics. The group also includes a tholeiitic to calc-alkaline, Fe-rich basal-

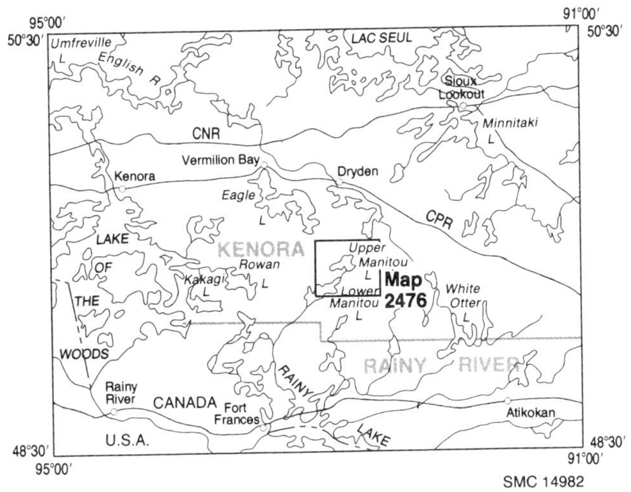

***Figure 1**—Key map showing location of the Manitou Lakes area.*

tic to rhyolitic sub-group that is laterally equivalent to these seven formations, and a gabbro body that is probably subvolcanic.

Correlation is made across the Manitou Straits Fault between the Wapageisi Lake and Blanchard Lake groups, and between the Manitou and Upper Manitou Lake groups, but not between the Boyer Lake and Pincher Lake groups. The Mosher Bay – Washeibemaga Lake Fault is a thrust fault along which the Boyer Lake group has been emplaced above the Manitou group.

Stratigraphic units northwest of the Manitou Straits Fault are tightly folded about the Manitou Anticline. Southeast of the Manitou Straits Fault, stratigraphic units south of the Mosher Bay – Washeibemaga Lake Fault face homoclinally north, whereas the Boyer Lake group is folded about the Kamanatogama Lake Syncline.

Gold mineralization is interpreted to be in major part epigenetic and to have been generated by metamorphic secretion from mafic volcanic rocks and felsic subvolcanic bodies during epizonal intrusion and tectonism. Base metal sulphide mineralization is interpreted to be stratigraphically controlled, within mafic metavolcanics and sills. Molybdenum mineralization occurs in residual hydrous phases of the Atikwa Batholith. Iron mineralization is of two types, sedimentary and magmatic. The former is considered to be a pelagic accumulation of a chemical precipitate within clastic sediments, the latter a differentiated phase of a gabbroic intrusion.

Geology of the Manitou Lakes Area
District of Kenora
(Stratigraphy and Petrochemistry)

by
C.E. Blackburn[1]

INTRODUCTION

The Manitou Lakes map area lies between Latitudes 49°15′ and 49°30′N and Longitudes 92°30′ and 93°00′W, in the District of Kenora. The area covers 1016 km^2, being 36.4 km wide by 27.9 km long. Mosher Bay, at the centre of the map-area, is 45 km south of Dryden, the nearest town, situated on the Trans-Canada Highway.

The Manitou Lakes, comprising Upper Manitou Lake and Lower Manitou Lake, were the scene of considerable gold prospecting and mining activity during the period 1895 to 1912, and again in the 1930s (Thompson 1933, 1938). Mining activity, during the same periods, was concentrated close to the north end of Trafalgar Bay of Upper Manitou Lake, and the community that developed came to be known as Gold Rock. Of the numerous occurrences and prospects near Gold Rock, the only three to come to production were the Big Master Mine during the period 1902 to 1905 and again in 1942 and 1943, the Laurentian Mine during the period 1906 to 1909, and the Elora (or Jubilee) Mine in 1936 to 1937, and in 1939. Other occurrences and prospects were scattered around the shores of the Manitou Lakes; only one of these came to production, the Twentieth Century Mine, near the south shore of Upper Manitou Lake, during the period 1902 to 1903. Recorded production from these mines is as follows (Ferguson *et al.* 1971).

	Years	Gold (oz)	Silver (oz)	Total Value (dollars)	Ore Milled (tons)
Big Master Mine	1902-1903 1905, 1942-43	2565[1]	184[2]	75,115[2]	14,470
Laurentian Mine	1906-1909	8143	–[3]	141,140	19,950
Elora Mine	1936, 1937, 1939	1370	296	49,017	13,766
Twentieth Century Mine	1902-1903	–[4]	–[3]	43,586	8,688

[1]No figures for 1902.
[2]No silver values for 1902, 1903, 1905.
[3]No production recorded.
[4]No figures available.

[1]Geologist, Precambrian Geology Section, Ontario Geological Survey, Ministry of Natural Resources, Toronto. Published with the permission of E.G. Pye, Director, Ontario Geological Survey. Manuscript approved for publication March 13, 1980, by V.G. Milne, Chief Geologist, Ontario Geological Survey.

No further mineral production has been reported from the map-area to date. During the 1960s attention turned to geophysical methods of exploration for base metals, and the Manitou Lakes have received attention by a number of companies in this regard (Blackburn 1976, 1979a, b).

Prior to the 1972 field season, when the author commenced investigations in the map-area, the only currently available published geological map covering the Manitou Lakes area was that of Thomson (1933) who surveyed an area of approximately 1300 km^2 at a scale of 1:63,360 (1 inch to 1 mile) during the summer of 1932.

Present Geological Survey And Compilation

The field work for the present project was carried out by the author and his assistant during the summer of 1976. The entire area had been mapped previously by the author and his assistants over the period 1972 to 1975 (Blackburn 1976, 1979 a, b), on field maps prepared at a scale of 1:15,840 (1 inch to ¼ mile) for final publication at 1:31,680 (1 inch to ½ mile). In 1976 field work consisted of re-examining critical outcrops, sections, and contacts for the purpose of synoptic compilation. In addition a number of sections were sampled, at right angles to strike of stratigraphy, for total rock chemical analysis of major and minor elements.

The synoptic map was compiled at a scale of 1:50,000 from photo-reductions of the four detailed maps, with amendments following the 1976 field season's work.

Access

Since the preparation of the detailed reports (Blackburn 1976, 1979a, b), Highway 502 has been completed, linking Highway 11 east of Rainy Lake with Highway 17 at Dryden. The highway passes through the east side of the map-area. Within the map-area access points are provided from this highway into Mosher Bay and Rattlesnake Lake via a canoe route, Scattergood Lake at its southeast end, and Meggisi Lake via the Trout River. Access to the rest of the area remains as described in the detailed reports listed above.

Previous And Current Geological Work

The first notable geological survey of the Manitou Lakes was undertaken by William McInnes (1895, 1896, 1897) of the Geological Survey of Canada. McInnes' map (1902) shows considerable detail within the west and central parts of the present map-area.

Geologists and inspectors of the then Ontario Bureau of Mines made short visits to mines and prospects of the Manitou Lakes area during the period from about 1895 to 1913, when an active "gold rush" was in progress. A.P. Coleman (1894, 1896) and A.L. Parsons (1911, 1912) in the course of inspections for the bureau, made pertinent geological observations during this period. E.L. Bruce (1925) reported on mining developments in 1925, and on his own field geological observations in the Manitou Lakes area.

The second geological survey covering the Manitou Lakes area was made in 1932 by Thomson (1933). At that time an intensive search for gold deposits was being made in the older gold regions, and the Manitou Lakes were receiving some new attention. Thomson returned to Upper Manitou Lake in 1937, and reported specifically on developments at Gold Rock (Thomson 1938).

In 1936, F.J. Pettijohn (1937), while studying Archean sedimentation, carried out geological mapping within the present map-area at Mosher Bay and Uphill Lake. During the 1963 and 1968 field seasons, A.M. Goodwin (1965, 1970) carried out stratigraphic studies for the then Ontario Department of Mines, allied with systematic sampling of metavolcanic units, over the Kenora-Fort Frances area, and included the Manitou Lakes in this work.

Over the period 1972 to 1975, the present author conducted geological surveys over the following areas: 1972, Lower Manitou-Uphill Lakes (Blackburn 1976); 1973, Upper Manitou Lake (Blackburn 1979a); 1974, 1975, Boyer Lake-Meggisi Lake (Blackburn 1979b). The present study is a synopsis of these reports.

Two recent Ph.D. studies concern the present map-area. A study by P.R. Teal of the stratigraphy, sedimentology, and associated volcanology of the Manitou Group was contained entirely within the area (Teal 1979). Preliminary results were given by Teal and Walker (1977). D. Birk made a Rb/Sr geochronology study of granitoid plutons in a portion of northwest Ontario (Birk 1978). The Taylor Lake and Scattergood Lake Stocks, contained in the present map-area, were included in Birk's study. Preliminary results were given in Birk and McNutt (1977).

The present map-area was included in a special study, under the direction of N.F. Trowell and this author, of stratigraphy, structure, and place and timing of mineral deposits in the Savant-Crow Lakes region (Trowell *et al.* 1980). A continuing geochronologic study run in conjunction with this latter project, under the direction of D.W. Davis, included samples from the present map-area. Preliminary results were given in Davis *et al.* (1980).

Acknowledgments

The author was ably assisted in the field during the 1976 field season by junior assistant George Gorzynski. Thanks are due to Al and Joanne Faloon of Green Island Lodge for courtesies and hospitality extended to the field crew during the 1976 field season.

GENERAL GEOLOGY

Isotopic ages obtained by Davis *et al.* (1980) from the present map-area indicate bedrock to be of Early Precambrian age, except for younger, cross-cutting, predominantly northwest-trending diabase dikes, which by analogy with isotopic dates obtained elsewhere in this part of northwestern Ontario (Wanless 1970), are considered to be of Middle to Late Precambrian age.

The map-area includes the west and central part of the Manitou-Stormy Lakes metavolcanic-metasedimentary belt, an arcuate structure some 20 km wide and 80 km long, extending from Lower Manitou Lake on the west to Bending Lake on the east, and tapering at either end. To the southwest, the belt is continuous with metavolcanics and metasediments that can be traced uninterruptedly via Kakagi Lake to Lake of the Woods, while to the north the belt joins with metavolcanics that extend northward towards Dryden (Blackburn *et al.* 1981). The Early Precambrian rock succession consists of a number of thick volcanic sequences, consisting of mafic to felsic flow and pyroclastic rocks and minor clastic and chemical sedimentary rocks, and intercalated sedimentary sequences, predominantly clastic, but with some chemical sedimentary rocks. Mafic to felsic rocks of batholithic, stock, and sill-like form intrude these supracrustal sequences at various levels.

Reconnaissance mapping of the Manitou-Stormy Lakes belt by Thomson (1933) led him to postulate that a lithologically distinct group of metasediments, that he named the Manitou "Series", lay above all the metavolcanics in the area he mapped. He thereby implied the presence of a major syncline along the central axis of the belt. Goodwin (1965, 1970), on the basis of a relatively small number of top determinations, interpolated syncline-anticline couples flanking Thomson's major syncline on both sides.

Recognition of Thomson's central synclinorial axis is complicated by a major fault zone, the Manitou Straits Fault, which in the western portion of the Manitou-Stormy Lakes belt lies along the supposed axis and forms the northwest border of Thomson's Manitou

TABLE 1. LITHOLOGIC UNITS FOR THE MANITOU LAKES AREA.

CENOZOIC
QUATERNARY
PLEISTOCENE AND RECENT
Clay, sand, gravel, boulders, muck.

Unconformity

MIDDLE TO LATE PRECAMBRIAN (PROTEROZOIC)
MAFIC INTRUSIVE ROCKS
Diabase.

Intrusive Contact

EARLY PRECAMBRIAN (ARCHEAN)
POST TECTONIC INTRUSIVE ROCKS
FELSIC PLUTONIC ROCKS
Porphyritic, seriate and equigranular hornblende-biotite granodiorite and quartz monzonite; pegmatite, aplite.

Intrusive and Gradational Contact

INTERMEDIATE PLUTONIC ROCKS
Hornblende monzonite; hornblende diorite and syenodiorite.

Intrusive Contact

LATE TECTONIC INTRUSIVE ROCKS
FELSIC PLUTONIC ROCKS
Equigranular, seriate and porphyritic biotite quartz monzonite and granodiorite; quartz monzonite and granodiorite gneiss; pegmatite, aplite.

Intrusive and Gradational Contact

INTERMEDIATE PLUTONIC ROCKS
Hornblende and biotite-hornblende diorite and quartz diorite.

Intrusive Contact

EARLY TECTONIC INTRUSIVE ROCKS
FELSIC PLUTONIC ROCKS
Hornblende-biotite quartz monzonite and granodiorite; quartz monzonite and granodiorite gneiss; fine-grained granitic rock; pegmatite, aplite.

Intrusive Contact

EARLY TO LATE TECTONIC INTRUSIVE ROCKS
FELSIC HYPABYSSAL ROCKS[a]
Quartz-feldspar and feldspar-quartz porphyry; felsite; granophyre and granitic rocks; sericitized and sheared felsic hypabyssal rocks; carbonatized felsic hypabyssal rocks; aplite.

Intrusive Contact

INTERMEDIATE HYPABYSSAL ROCKS
Microgranodiorite and micro quartz diorite porphyry.

Intrusive Contact

METAMORPHOSED MAFIC AND ULTRAMAFIC INTRUSIVE ROCKS[a]
Gabbro; troctolitic gabbro; diabase; pyroxenite and pyroxenitic gabbro; peridotite; lamprophyre; granophyre[b].

Intrusive Contact

METASEDIMENTS
CHEMICAL METASEDIMENTS
Magnetite ironstone; chert.
CLASTIC METASEDIMENTS
Conglomerate; sandstone, mudstone; sericite schist.
METAVOLCANICS
ALKALIC MAFIC METAVOLCANICS
Hornblende-feldspar-phyric flows; amygdaloidal flows; volcanic breccia.
SUBALKALIC FELSIC METAVOLCANICS
Tuff, lapilli-tuff; tuff-breccia; flows; quartz-feldspar porphyry; sericite-chlorite schist.
SUBALKALIC INTERMEDIATE METAVOLCANICS
Tuff, lapilli-tuff; tuff-breccia; quartz-feldspar-biotite schist and gneiss; chlorite-sericite schist.
SUBALKALIC MAFIC METAVOLCANICS
Medium- to fine-grained flows; coarse-grained flows (gabbroic); pillowed flows; plagioclase-phyric flows; plagioclase-phyric pillowed flows; autoclastic breccia; plagioclase-phyric coarse-grained flows; amygdaloidal flows; variolitic flows; amphibolite, amphibolitic migmatite; chloritic schist; carbonatized flows; tuff, lapilli-tuff; tuff-breccia.

[a]Rocks grouped under these headings are not all the same age.
[b]Felsic differentiate phase related to gabbroic rocks.

"Series" (Thomson 1933; Goodwin 1965, 1970; Blackburn 1976, 1979a, b; Blackburn *et al.* 1981).

During the detailed mapping of the west part of the Manitou Lakes area, the author (Blackburn 1976, 1979a) found evidence to support one only of Goodwin's interpolated anticlines (Goodwin 1965, 1970), the Manitou Anticline (Goodwin 1970), to the northwest of the Manitou Straits Fault. On the southeast side of the fault, during mapping of the south part of the present area (Blackburn 1976, 1979b) no evidence was found for the syncline-anticline couple interpreted by Goodwin to flank Thomson's major syncline on the southeast side: on the contrary, the sequence was found to face homoclinally north. On this same southeast side of the Manitou Straits Fault, in the northeast part of the present map-area, the author (Blackburn 1979b) found that sequences had been folded about an east-trending syncline that is truncated by the major fault: thus, if a major synclinorial axis exists in the Manitou-Stormy Lakes belt, it does not lie within the Manitou "Series" of Thomson, but lies north of these metasediments.

Discussion of lithologies (Table 1) and rock sequences have been given in reports by the present author (Blackburn 1976, 1979a, b). In the following sections, summaries only are given of physical characteristics of predominant lithologies. The reader is referred to the reports on the detailed mapping for further details. Chemical characteristics are discussed following an account of the stratigraphy.

Early Precambrian

METAVOLCANICS

Criteria for subdivision of the metavolcanics were consistent throughout the detailed mapping (Blackburn 1976, 1979a, b). It is possible to subdivide the metavolcanics into broad categories based on a combination of texture, structure and mineralogy. Textural and structural criteria were used to assign rocks to either flow or pyroclastic categories, while a combination of such factors as colour of the fresh surface, colour index, hardness, weathering colour, and visible quartz content were used to assign metavolcanics to mafic, intermediate, or felsic categories. For pyroclastic rocks composed of mixed volcanic rock types, and also various matrix compositions, the overall colour index estimated by inspection of clasts and matrix was used to categorize the rock.

During the detailed mapping it became evident that most of the metavolcanics could be classified as either mafic (colour index greater than 35) flow rocks, or intermediate to felsic (colour index less than 35 and ranging to less than 15) pyroclastic rocks. However, in parts of the map-area, pyroclastic rocks were found to be mafic, while some flows were found to be felsic. Partial and complete chemical analyses of selected metavolcanics from the detailed map-areas, when plotted on alkalies versus silica diagrams according to the volcanic rock classification of Irvine and Baragar (1971), reveal that subalkaline rocks, ranging from basalt to rhyolite, predominate, but that a distinctive suite of alkaline rocks, of trachybasalt affinity, occupy a distinct stratigraphic level.

On the detailed maps, felsic metavolcanics were grouped with intermediate metavolcanics. In the present synoptic survey they have been distinguished as a separate lithology (Map 2476, back pocket).

Subalkalic Mafic Metavolcanics

Rocks assigned to this broad class probably make up more than two-thirds of the volcanic piles within the Manitou Lakes area. The majority are flow rocks.

Evidence of their submarine extrusion is given by ubiquitous pillows. The form of the pillows varies considerably, from small, equant and bun-shaped, to large, elongate, and irregular. Abundance of inter-pillow material and thickness of rims are also variable. Pillows are variably composed of massive basalt or contain amygdules, feldspar phenocrysts, and, at Boyer Lake, variolites. Some pillows contain epidotic centres. In places pillow breccia occurs with hyaloclastite. All of these features can be used to suggest depth and environment of emplacement, and will be discussed later in the report.

Plagioclase-phyric flows occur at a number of stratigraphic levels, in some cases as distinctive marker units which can be mapped up to 10 km along strike, and which are on the order of 1000 m thick. Some of these thick units may be shallow, synvolcanic sills, but the presence of pillows within others confirms their flow origin. At other places plagioclase phenocrysts are ubiquitous throughout a sequence of flows, either as small single crystals, or in cumulophyric aggregates.

Coarse crystalline, gabbroic, phases occur as thick stratigraphic units, traceable along strike over many kilometres, and may be either thick flows, or sills, but essentially synvolcanic. They are distinctly different from other definite gabbro sills in that they lack ophitic texture, and grade into basaltic flows without distinct contacts. Presence of pillows within some of these coarse phases suggests their essentially shallow, synvolcanic, emplacement.

The subalkalic mafic metavolcanics occur either in very thick, rather monotonous, sequences, thousands of metres thick, the maximum estimated recorded thickness for such a sequence being about 6700 m for any one sequence, or in association with subalkalic intermediate and felsic units, where they form narrower flows, frequently discontinuous along strike. Chemically, the thick sequences tend to be tholeiitic, and the narrow flows tend to be calc-alkaline. Also the thick sequences mostly show evidence of deep to moderate depths of subaqueous emplacement, whereas those interbedded with intermediate and felsic rocks that are mostly pyroclastic tend to be highly vesicular and brecciated by seawater action, indicative of a shallow, but subaqueous, extrusion.

Towards the margins of the supracrustal belt, near contacts with granitic rocks of the Atikwa Batholith in the northwest and the Irene-Eltrut Lakes Batholithic Complex in the southeast, amphibolite grade of metamorphism prevails, and textures and structures are obscured, both by metamorphism and by deformation. Xenoliths of engulfed mafic rocks occur well within the granitic batholiths, as recognised by remnant pillow structures, feldspar-phyric remnants, and other features.

At the edge of the Atikwa Batholith, voluminous dioritic and quartz dioritic rock occupies a transition zone between the batholith and the volcanic belt. The rocks have been interpreted by the author (Blackburn 1976, 1979a) to be of hybrid origin, derived by assimilation, metasomatism, or partial melting of primary volcanic and gabbroic rocks by invading granitic magmas of the batholith. Heimlich (1971) came to the conclusion that similar rocks at Atikwa Lake termed hornblende tonalite by him, but similar petrographically and modally to quartz diorite and diorite of the present area, were derived by assimilation of mafic volcanic rock by invading acidic magma of the Atikwa Batholith. He suggested that the magma "was contaminated because of an exchange of chemical components between the magma and the greenstone inclusions. This exchange was facilitated by the presence of volatiles, largely water, supplied by the intruding magma" (Heimlich 1971, p.12). Dioritic rocks are common marginal to the Atikwa Batholith (e.g. Eagle Lake area: Moorhouse 1941; Dryden-Wabigoon area: Satterly 1943; Atikwa Lake area: Davies 1973; Populus Lake area: Davies and Watowich 1958; and the Manitou Lakes: Blackburn 1976, 1979a), but are not found marginal to the Irene-Eltrut Lakes Batholithic Complex.

Subalkalic Intermediate Metavolcanics

The majority of pyroclastic rocks in the area are of subalkalic intermediate composition. Flows of this composition have not been recognised in association with pyroclastic rocks, though they may be present, to date undetected, within the more mafic flow sequences.

During the detailed mapping, genetic terminology was avoided. The fragment size and mixture classification of Fisher (1966) was employed. An attempt was made to distinguish between heterolithic and monolithic deposits in the lapilli-tuff, tuff-breccia and volcanic breccia mixture ranges, though this was not indicated in the published maps. For all mixture ranges considerable difficulty was found in distinguishing between subaerial and subaqueous deposits, and also between pyroclastic rocks and epiclastic sedimentary rocks. These problems were compounded towards the northwest side of the supracrustal belt, marginal to the Atikwa Batholith, and also within xenoliths in the granitic rocks of the batholith, where amphibolite facies metamorphism prevails.

On the basis of associations observed during the detailed mapping, the author proposes that pyroclastic rocks within the map-area are probably of subaqueous origin if they 1) are heterolithic, 2) are interbedded with pillowed flows, 3) contain sandy interbeds that are lensoid, and may be cross laminated, with evidence of scouring, and 4) contain graded beds in tuff-size material, with possible associated load casts and flame structures.

They are, conversely, probably subaerial if they lack features listed under 2, 3, and 4 above, and if they 1) are monolithic, 2) contain bedding in which the bedding planes are diffuse, and grading is in both directions, and 3) contain bombs in tuff-size material.

All of these features are present in outcrops in the Manitou Lakes area. Other textures and structures are less diagnostic of environment.

Subalkalic Felsic Metavolcanics

During the detailed mapping felsic metavolcanics were distinguished from intermediate types, but were placed under the same lithologic numerical code. In the present report, subsequent compilation has shown that these rocks define precise units within the volcanic edifices, so that now they have been given separate status from the intermediate metavolcanics.

Felsic metavolcanics are of flow, pyroclastic, and autoclastic origin. Flows, of dacitic to rhyolitic composition, occur at a number of localities, scattered through both pyroclastic sequences and within mafic flow sequences with pyroclastic intercalated units. One distinct association is with hypabyssal quartz-feldspar porphyry intrusions at Sunshine Lake and at Washeibemaga and Thundercloud Lakes.

Pyroclastic felsic rocks are dominantly fine grained, in the tuff, lapilli-tuff, and lapillistone mixture categories. In particular, tuffaceous units occur as distinctive marker beds, particularly at Upper Manitou Lake where they can partially be traced around the Manitou Anticline. Coarser felsic pyroclastic rocks occur on the east limb of the Manitou Anticline, in the vicinity of the Manitou Straits.

Felsic autoclastic breccia occurs at Washeibemaga Lake, associated with the hypabyssal Thundercloud porphyry. The association of quartz-feldspar porphyry with an overlying brecciated phase and rhyolitic to dacitic flows indicates presence of a felsic vent that has been plugged by its own magma. The Sunshine Lake porphyry and associated dacitic to rhyolitic flows suggest a similar structure.

Alkalic Mafic Metavolcanics

A minor, but nonetheless very significant, class of volcanic rock in the area is of alkalic nature. The rocks are predominantly flows, though some pyroclastics appear to be present. Characteristically, the flows are coarsely porphyritic (feldspar and amphibole phenocrysts) and commonly amygdaloidal. The flows are petrochemically similar to a conformable series of mafic sills that occur stratigraphically just beneath them, and the author when mapping in the Lower Manitou-Uphill Lakes area (Blackburn 1976) interpreted all of these units to be sills, though noted that to the northeast the upper unit appeared to pass into flows. Subsequent mapping (Blackburn 1979b) has led to the re-interpretation that the uppermost of the three bodies are flows, based upon high vesicularity and presence of breccia.

The alkalic metavolcanics have been interpreted by Teal (1979) to be subaerial, mostly due to negative evidence, such as lack of pillows. However, presence of rubbly breccia in a matrix of the same composition and texture, and lacking glassy or hyaloclastic phases, is also indicative of subaerial emplacement, so that the present author concurs with Teal.

METASEDIMENTS

Criteria for subdivision of the metasediments were consistent throughout the detailed mapping (Blackburn 1976, 1979a, b). Clastic metasediments were mapped under four groups: volcanic-clast conglomerate; polymictic conglomerate; sandstone; and siltstone, argillite, slatey argillite. Chemical metasediments were mapped under two groups: chert; and ironstone. Because of the scale of the survey it was not considered practical to further subdivide the sandstones during the field mapping: detailed studies of sedimentary petrography may enable subdivision.

Considerable difficulty was encountered in distinguishing coarse pyroclastics from epiclastic conglomerate with abundant volcanic clasts, into which they appear to grade in places. The following criteria, especially where found in conjunction, were used to classify deposits as volcanic-clast conglomerates: 1) presence of a minor proportion of clasts of undoubted non-volcanic origin, 2) sharp clast boundaries, and 3) presence of sandstone interbeds.

Polymictic conglomerate is composed of volcanic clasts plus abundant clasts of undoubted non-volcanic origin, such as granitoid, chert, and ironstone clasts. In practice a complete gradation was found between volcanic-clast and polymictic conglomerates, but polymictic conglomerates tend to be interbedded with sandstones and mudstones.

For present purposes, clastic metasediments are subdivided into conglomerate, and sandstone and mudstone, and chemical metasediments into magnetite ironstone and chert.

Clastic Metasediments

Clastic metasediments are by far the predominant metasediments in the area, and were early recognized as constituting a distinct stratigraphic unit, the Manitou "Series" (Thomson 1933).

Recent studies by P.R. Teal (Teal 1979; Teal and Walker 1977) have been directed toward determining the depositional environment of the clastic sediments of Thomson's Manitou "Series". Textures, gross structure and petrography have led Teal to the conclusion (Teal 1979) that facies of essentially two environments are represented, a terrestrial facies, and a submarine facies.

Conglomerates at Manitou Lakes show a wide range of structure and composition. Mapping by the author (Blackburn 1976, 1979b) showed that volcanic-clast conglomerates are prevalent in lower sections of the Manitou "Series", with an increase in exotic, nonvolcanic, clasts upwards. Quantitative studies by Teal (1979, Table 8, p.154) confirmed this generalization: maximum and minimum averaged values for the following components as defined by Teal, show the following trends from units near the base to units near the top of the epiclastic sequences: decrease in "basalt" clasts (53 percent to <1 percent); decrease in "porphyry" clasts (33 percent to <1 percent); increase in "granite" clasts (0 to 32 percent); increase in "sedimentary" clasts (0 to 14 percent); increase in "iron formation" clasts (0 to 10 percent); and increase in "matrix" (13 percent to 59 percent), where "matrix" was considered to be grains under 1 cm diameter.

Similarly, the detailed mapping (Blackburn 1976, 1979b) showed a progressive change in sandstone composition upwards in the Manitou "Series". Field observations suggested that sandstones of lower sequences contain abundant lithic, volcanic grains. Employing a simple classification scheme (Figure 2) they would be lithic arenite and wacke. Sandstones of higher sequences appeared to be more feldspar- and quartz-rich, with fewer lithic grains, and would be feldspathic to quartzose arenite and wacke. Quantitative studies by Teal (1979, Table 7, p.147) show trends along as well as across strike, and modify the above generalizations: maximum and minimum averaged values for the fol-

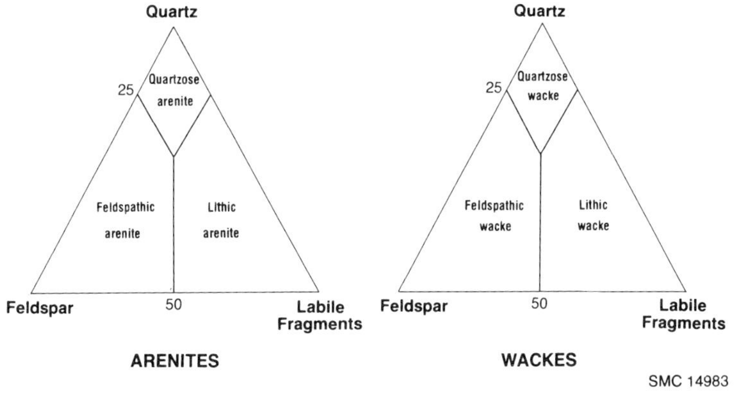

Figure 2*—Simplified sandstone classification, showing subdivision of framework components. Arenite is a sandstone with 0 to 15 percent fine matrix (less than 0.6 mm grain size); wacke is a sandstone with 15 to 75 percent fine matrix.*

lowing components show the following trends from units near the base to units near the top of the epiclastic sequence: decrease in volcanic rock fragments (52 to 12 percent); increase in quartz (1 to 15 percent); increase in "matrix" (33 to 60 percent), where "matrix" was considered to be material finer than 0.03 mm. Feldspar shows considerable changes both across and along strike, but ranges from 6 to 23 percent: the upward increase suggested by the author (Blackburn 1979b) was not confirmed. Average matrix values are consistently high, though a few values below 15 percent were recorded (Teal 1979, Appendix II, p.226-230). Despite the variations noted, most sandstones therefore appear to be lithic wacke, with minor feldspathic wacke, and even less feldspathic arenite.

Chemical Metasediments

Chemical metasediments constitute a minor but important class of metasediments in the Manitou Lakes area.

Magnetite ironstone[1], because of its supposed volcanic association, may constitute a valuable stratigraphic marker. It has been mapped at six localities, in all cases interbedded with clastic metasediments: at Beaverhead Island, at a point mid-way along Manitou Straits (Blackburn 1976), at three localities south of Mosher Bay, and north of Kennewapekko Creek at Washeibemaga Lake (Blackburn 1979b). All occur at approximately the same lithostratigraphic level, and probably represent a chronostratigraphic level.

Characteristically the magnetite ironstone is interbedded with sandstone and mudstone. By far its greatest development is at two of the six localities: at Beaverhead Island the combined ironstone-sandstone unit is 15 m wide by 800 m long; at Mosher Bay, the most westerly of the three occurrences, a mixed ironstone-mudstone sequence, is 15 m wide by more than 200 m long. The ironstone beds vary from less than 1 cm to a maximum of 30 cm in thickness, alternating with similar thicknesses of sandstone and mudstone. Minor red jasper lenses occur with some of the magnetite beds, especially at Beaverhead Island (Blackburn 1976). At Beaverhead Island and at the locality directly on the south shore of Mosher Bay the magnetite ironstone-sandstone units have been strongly folded.

A 15 to 30 m thick massive chert unit occurs in the southwest corner of the map-area, where it can be traced discontinuously a distance in excess of 5 km along strike in the vicinity of Meridian Bay and Holcroft Lake. The chert unit is associated with sandstone, mudstone, and minor conglomerate.

METAMORPHOSED MAFIC AND ULTRAMAFIC INTRUSIVE ROCKS

Mafic and ultramafic intrusive rocks constitute a variety of stock, sill and dike-like bodies of varying petrologic affinities in the area. They are discussed together because clear-cut distinctions between them are difficult to make. In addition, many rocks mapped as coarse mafic flows in the field may be in reality intrusive, as discussed under "Subalkalic Mafic Metavolcanics".

Southeast of the Manitou Straits Fault pyroxenitic to peridotitic intrusive rocks occur near the base of the metavolcanic sequence (Blackburn 1979b).

Two thick sills south of Sunshine Lake (Blackburn 1979b) are probably of different petrologic affinity and age. Their relationship with a quartz-feldspar porphyry stock suggests the more southerly, which is intruded by the stock, to be older than the more northerly, which appears to intrude the felsic porphyry stock.

[1] The term ironstone as used here is a chemical sedimentary rock that contains 33 percent or more of the common iron minerals by volume. This definition excludes other chemically precipitated sediments such as chert, and clastic sedimentary material, that are commonly interlayered with ironstone.

The lamprophyric to peridotitic nature of the northerly sill suggests its correlation with a sill at Cane Lake described as diabasic and porphyritic by Blackburn (1976, Table 5, samples 24 and 26). Lamprophyre dikes, many of which are too small to show on Map 2476 (back pocket) but which are indicated on Map 2320 (Blackburn 1976) and Map P.1188 (Blackburn 1979b), are probably of the same petrogenetic affinity as these latter sills (Blackburn 1976, 1979b). The lamprophyric and peridotitic rocks are probably sub-volcanic manifestations of the trachybasaltic flows and pyroclastics discussed under "Alkalic Mafic Metavolcanics". Their chemical similarities and differences are discussed under "Petrochemistry, Manitou Group").

A number of large mafic to ultramafic bodies intrude a thick sequence of subalkalic mafic metavolcanics that occupy an area southeast of the Manitou Straits Fault, but north of Mosher Bay and Washeibemaga Lake. These were previously referred to as the Boyer Lake, Mountdew Lake, Washeibemaga, and Mosher Bay gabbros (Blackburn 1979b). The Boyer Lake body has been shown (McMaster 1975) to be a fractionally crystallized and differentiated horizontally emplaced sill. Petrographically, variation is from gabbro to troctolitic gabbro to pyroxenitic gabbro and diabasic marginal phases. Insufficient study has been carried out on the other bodies to define distinct phases as at Boyer Lake.

Northwest of the Manitou Straits Fault, the Mitchell Lake Gabbro intrudes the metavolcanic sequence at the northwest edge of the belt. Petrographic descriptions and chemical analyses of four samples typical of various phases of this stock-like body (Blackburn 1979a) indicate it is gabbroic, but varies considerably, from leucocratic, andesine-rich, phases to melanocratic, hornblende-rich, phases. Considerable difficulty was found in placing the boundary between gabbro of the stock and diorite marginal to the Atikwa Batholith.

FELSIC AND INTERMEDIATE INTRUSIVE ROCKS

Hypabyssal Intrusions

Dikes, sills, and small stocks of porphyry, granophyre, and related microgranitic rocks, ranging in composition from dacitic to andesitic, intrude all levels of the volcanic and sedimentary pile. They are predominantly felsic, quartz-feldspar porphyries, with various proportions of phenocrysts and matrix. Feldspars are predominantly albitic plagioclase, with potassic feldspar occurring in minor amounts. Chemically, these porphyries are dacitic and rhyolitic (Blackburn 1979b; McMaster 1978). Details of their textures and mineralogy are discussed further by Blackburn (1976, 1979a,b) and McMaster (1978).

A distinctly different group of intermediate to mafic hypabyssal intrusive rocks occurs at Frenchman Island of Upper Manitou Lake. Blackburn (1979a) previously grouped them with the metavolcanics, but they are here given a separate heading, recognising their subvolcanic but essentially intrusive nature. The microgranodiorite and micro quartz diorite vary from porphyries, with quartz phenocrysts, to equigranular rocks. Chemically, they vary from dacitic through andesitic to basaltic (Blackburn 1979a). Textures and mineralogy are discussed further by Blackburn (1979a), who also pointed out their close similarity to adjacent coarse pyroclastics at Upper Manitou Lake, and suggested the Frenchman Island body to have been emplaced in the vent for these pyroclastics.

Batholithic Granitic Complexes*

The volcano-sedimentary belt is bounded to the northwest and southeast by granitoid rocks of two batholithic complexes that extend well outside the map-area. These respec-

*Note: On the map, the granitic rocks have been subdivided into units which comform with other compilation maps in this region. In detail, this leads to some inconsistencies with phases in the report. The map-unit(s) corresponding to each phase are given in Table 2 so that the various phases can be identified on the map.

TABLE 2. COMPARATIVE TABLE OF PETROGRAPHY AND TECTONIC SETTING OF GRANITOID ROCKS IN THE MANITOU LAKES AREA.

	BATHOLITHIC GRANITIC COMPLEXES				
	ATIKWA BATHOLITH[1]			MEGGISI PLUTON[2]	
	PHASE A (Map-Unit 12a)	PHASE B (Map-Unit 12 b)	PHASE C (Map-Unit 11a)	PHASE A (Map-Unit 10a)	PHASE B (Map-Unit 12a)
Fabric	Equigranular	Porphyritic	Equigranular	Equigranular	Seriate
Composition	Granodiorite	Quartz monzonite	Diorite, quartz diorite	Granodiorite	Quartz monzonite
Mineralogy (accessories in brackets)	Q+P+K+E±B (±S±A±Z±O)	Q+P+K+E±B (±S±A±Z±O)	H+P±B±Q±A±E (±S±Z±O)	Q+P+E+B+H+E (±S±A±O±C)	Q+P+K+B+E (±S±O±A)
Plagioclase					
a) Zoning	Normal, weak	Normal, weak	Normal, strong	Normal and oscillatory	Normal and oscillatory
b) Composition	Calcic oligoclase	Calcic oligoclase	Andesine cores; oligoclase rims	Andesine & oligoclase	Sodic oligoclase
c) Twinning	Albite, Carlsbad, Pericline	Albite, Carlsbad, Pericline	Albite, Carlsbad, Pericline	Albite, Carlsbad	Yes
Potassic feldspar	Microcline	Microcline	None	Microcline	Microcline
Dominant mafic minerals	Biotite	Biotite	Hornblende	Biotite (± hornblende)	Biotite
Tectonic setting	Late tectonic	Late tectonic	Late tectonic	Early tectonic	Late tectonic

INTRABELT PLUTONS				
SCATTERGOOD LAKE STOCK[3]	CARLETON LAKE STOCK[4]	TAYLOR LAKE STOCK[5]		
		PHASE A	PHASE B	PHASE C
(Map-Unit 14a)	(Map-Unit 14a, b)	(Map-Unit 13a, b)	(Map-Unit 14a)	(Map-Unit 14b)
Porphyritic to seriate	Seriate	Equigranular	Seriate to porphyritic	Equigranular
Quartz monzonite	Quartz monzonite	Monzonite, syenodiorite	Quartz monzonite	Granodiorite
P+Q+K+H+E (±B±S±A±Z±C±M)	P+Q+K+B (±S±O±E±C)	Pn+H+K+P±B±Q (±E±S±A±O)	H+Q+P+K±B (±Pn±E±S±O)	P+Q+B+K±H (±E±S±C±M±O)
Normal Oligoclase	Normal Oligoclase	No ? Altered	Yes Altered	No ? Altered
Yes	Yes	Untwinned?	Yes	Untwinned?
Microcline	Microcline	Microcline	Microcline	Microcline
Hornblende	Biotite	Pyroxene + hornblende	Hornblende ± biotite	Biotite
Post tectonic	Post tectonic	Post tectonic	Post tectonic	Post tectonic

Q = quartz; P = plagioclase; K = potassic feldspar; E = epidote; B = biotite; H = hornblende; Pn = pyroxene; S = sphene; A = apatite; Z = zircon; O = opaques; C = chlorite; M = muscovite (sericite).
[1]Data from: Blackburn (1976, 1979a).
[2]Data from: Blackburn (1979b), Sabag (1979).
[3]Data from: Blackburn (1976, 1979b), Birk (1978).
[4]Data from: Blackburn (1976) and unpublished notes.
[5]Data from: Blackburn (1979b), Pichette (1976).

tively are the Atikwa Batholith, the portion within the map-area being termed the Kaminni Lake lobe by Blackburn (1976), and the Irene-Eltrut Lakes Batholithic Complex, the portion within the map-area being termed the Meggisi Lake lobe by Blackburn (1979b). Modal analyses (Blackburn 1979a,b; Sabag 1979) of samples from both bodies revealed wide compositional ranges, but essentially quartz monzonite to granodiorite, employing a classification scheme comparable to that of Bateman *et al.* (1963). Mafic content is generally low. Plagioclase is commonly zoned, both normally and rhythmically, with compositions from andesine to albite recorded, and the potassic feldspar is microcline (Blackburn 1979a,b; Sabag 1979).

Northwest of the supracrustal belt, Blackburn (1976, 1979a) delineated essentially three phases within the Atikwa Batholith (Table 2), equigranular granodioritic (phase A), porphyritic quartz monzonite (phase B), and equigranular diorite to quartz diorite (phase C). Blackburn (1979a) stated that all three phases appear to be gradational into each other. On the peninsula at the north end of Big Manitumeig Lake, field checking, in 1976 of the contact between phase A and a large body of phase B confirmed that, although the contact may be abrupt, no cross-cutting relationships can be observed. It therefore appears that phases A and B are coeval.

Southeast of the supracrustal belt, Sabag (1979) has delineated essentially two phases within the Meggisi Pluton, as discussed by Blackburn (1979b). An early equigranular granodiorite phase (Table 2, phase A) is intruded by a later seriate quartz monzonite phase (Table 2, phase B). During field checking at and near Meggisi Lake in 1976, the present author observed the contact at four localities. At two of these, one on the east shore of the north arm of Meggisi Lake, the other on the east shore of a lake west of here, the contact was observed to be sharp, but age relationships could not be established with certainty. At the contact on Meggisi Lake, phase A was found to be rich in mafic inclusions, a feature noted by Sabag (1979) throughout the body, right up to the contact, whereas the seriate phase is devoid of inclusions. At a third locality, on the west shore of the same lake west of the north arm of Meggisi Lake, the seriate phase was observed to cross-cut and intrude in veinlets the equigranular phase. Right at the contact the equigranular phase was observed to have a gneissic fabric, and to contain large mafic blocks in the form of an intrusion breccia. The seriate phase on the other hand is devoid of gneissosity and mafic inclusions. At a fourth locality on a lake immediately east of the east end of the Meggisi Lake, the contact was observed to be a lit-par-lit intrusion, with veinlets of the seriate phase intruding the equigranular phase. Both here and at the contact on the east shore of the small lake west of the north arm of Meggisi Lake the regional foliation was observed to cross-cut the contact between the two phases, implying its development after the emplacement of both phases.

As noted previously under "Subalkalic Mafic Metavolcanics", at the periphery of the Atikwa Batholith, dioritic phases are abundant. In contrast to the Atikwa Batholith, no quartz diorite or diorite phase is found at Meggisi Lake.

Intrabelt Plutons

Three plutons that clearly post-date emplacement and deformation of the supracrustal assemblage intrude the volcano-sedimentary belt. Two of these, the Scattergood Lake and Taylor Lake Stocks, are large, on the order of 30 to 40 km^2 in area; the third, the Carleton Lake Stock, is considerably smaller at about 3 km^2.

The Scattergood Lake Stock and the Carleton Lake Stock are similar bodies, of seriate to porphyritic, quartz monzonite to granodiorite (Blackburn 1976, 1979ab). Birk (1978) confirmed the author's previous (Blackburn 1979b) view that the Scattergood Lake Stock is remarkably homogeneous, both from modal analyses and chemical analyses. Only one chemical analysis is available (Blackburn 1976) but the Carleton Lake Stock appears to be essentially identical chemically as well as mineralogically to the Scattergood Lake Stock.

The Taylor Lake Stock is composed of a number of different phases ranging in composition from syenodiorite to monzonite to quartz monzonite to granodiorite. Its petrography is discussed by Blackburn (1979b), who summarized results of a study by Pichette (1976). On the present map (back pocket) a simpler subdivision is employed but in Table 2 a division into three essential phases is shown, in order to show the range in petrography. Pichette (1976) concluded that the stock is a composite intrusion of fractions from a single magma, with pulses intruded in progressively more acidic order. However, across the stock, there is a 5 km discontinuity in a distinctive basalt flow unit (formation 4 of the Wapageisi Lake group), and this suggests to the author that assimilation of mafic volcanic rocks may account for presence of mafic phases at the periphery of the stock.

Middle to Late Precambrian

DIABASE DIKES

In the Kenora-Fort Frances region, diabase dikes, because of their consistent cross-cutting relationships, have been generally considered to be considerably younger than all other rock formations. As is evident on the compilation map (Blackburn *et al.* 1981), these dikes consistently trend northwesterly, apart from the present map-area where a number of diabase dikes trend northeasterly. Petrographically they are similar to those of the northwesterly set in the area (Blackburn 1976, 1979b). As noted by the author (Blackburn 1979b), alteration is present to some degree in most of the diabase dikes, though the characteristic diabasic texture and brown-weathering colour makes tham all distinctive. No consistent relationship has been noted between orientation and degree of alteration, so the author considers that all the dikes are probably of the same generation.

STRATIGRAPHY OF THE MANITOU-STORMY LAKES VOLCANO-SEDIMENTARY BELT

Following reconnaissance field mapping in 1932, Thomson (1933) made a number of implications regarding the stratigraphy of the Manitou-Stormy Lakes area in general. He applied the name "Manitou Series" to "a well-defined band of sedimentary rocks .... The belt averages about 2 miles in width and is divided into two parts, which are on the same strike but separated by a mass of intrusive granite." (Thomson 1933, p.13). It appears that this correlation of the two parts was made partly on the basis of lithology and partly on the basis of the two parts lying above the same volcanic sequence. However, since it was implied rather than proven that all the volcanic rocks were of the same age (i.e. Keewatin) and that the structure of the belt was a synclinorium, then the sedimentary rocks, being in the centre of the synclinorium, by definition were correlatable, and of the same age (i.e. Timiskaming). Subsequent detailed mapping (Blackburn 1976, 1979a, b) has demonstrated that the volcanic rocks are not all of the same age, and that the structure is considerably more complex than that implied by Thomson. Although the sequences that constitute Thomson's Manitou "Series" may well be of the same age, their physical separation by the granitic intrusion necessitates assigning them to two separate lithostratigraphic units.

Teal and Walker (1977) have applied the term Manitou Group to that portion of Thomson's Manitou "Series" west of the Taylor Lake Stock. Although they do not discuss the matter, by implication they exclude from the Manitou Group that portion east of the stock. This procedure accords with that adopted here. Teal (1979) has adopted the same terminology. Teal (1979) has suggested formation names for sub-units of the Manitou Group, and because the author mostly concurs with Teal's subdivisions those names have been adopted here.

The names Sagenak conglomerate and Dark Horse conglomerate were applied by Bertholf (1946) to what he considered to be two distinct sedimentary units outcropping within the present area at Washeibemaga Lake. Subsequent mapping at Washeibemaga Lake by the author (Blackburn 1979b) and McMaster (1978) has shown that a simple two-fold division in this vicinity is not entirely justifiable, as discussed below under "Stormy Lake Group".

No other stratigraphic precedents have been set in the area. However, during detailed mapping, the author (Blackburn 1976, 1979a,b) discussed rock sequences by assigning names taken from local geographic features, pointing out that these names were not to be construed as formational names. Reference to descriptions of these rock sequences is given in Table 3. Some of those same names are now applied to formation, group, and sub-group categories described here[1].

The Manitou Straits Fault is a natural but not insurmountable barrier to correlation. Lithologic successions can be recognised within supracrustal and subvolcanic rocks on either side of the fault (Figure 3, Chart A, back pocket). It is therefore a convenient means by which to subdivide the synoptic area for purposes of description of stratigraphic units.

Southeast of the Manitou Straits Fault

Supracrustal and subvolcanic rocks southeast of the Manitou Straits Fault are here assigned to four lithostratigraphic groups. Of these the name of one, the Manitou group, has been proposed by Teal and Walker (1977), and is used by them and Teal (1979) to encompass the same formations as discussed here below.

Abundant top determinations, in both volcanic and sedimentary rocks, show that within three of the groups, the Wapageisi Lake, the Manitou, and the Stormy Lake, sequences face homoclinally northward. The Boyer Lake group, that lies north of a distinct east-trending break, the Mosher Bay – Washiebemaga Lake Fault, is shown on the basis of top determinations in metavolcanics to be folded about an east-trending axis.

The four groups are discussed below in structural order, from lowest to uppermost, noting that the Manitou and Stormy Lake groups are at the same structural level.

WAPAGEISI LAKE GROUP

This group is named after a large lake east of the present map-area. Reconnaissance mapping by Thomson (1933), supplemented by further checking by the present author during field work for a regional correlation project (Trowell *et al.* 1980), showed that the lithologies and sequences at Wapageisi Lake are continuous along strike with those discussed here.

At the present level of mapping the Wapageisi Lake group is divisible into a lower sub-group and two upper formations. Formations can tentatively be defined within the sub-group, but because of lack of suitable toponyms, are assigned numbers rather than names.

[1]In this report, formations, sub-groups, and groups are described as lithostratigraphic (rock-stratigraphic) units, according to precepts of the Code of Stratigraphic Nomenclature (ACSN 1961). Because not all of the conditions for establishing formal rock-stratigraphic units (ACSN 1961, Article 13, p.653) have been met in this report, all units are of informal status.

TABLE 3. CORRELATION OF STRATIGRAPHIC UNITS WITH LITHOLOGIC UNITS IN PREVIOUS REPORTS BY THE AUTHOR.

Stratigraphic Unit	Reference	Page Nos.	Lithologic Unit in Reference
Starshine Lake sub-group	Blackburn 1979b,		Starshine Lake Basalts
	1976,	p. 13 – 16	Beck Lake Basalts
Etta Lake formation	Blackburn 1976,	p. 16	Etta Lake Basalts
	Blackburn 1976,	p. 24 – 29	Intravolcanic Clastic and Chemical Sediments: Etta Lake Metasediments
Glass Bay formation	Blackburn 1976,	p. 16 – 17	Glass Bay Basalts
Cane Lake formation	Blackburn 1976,	p. 22 – 23	Cane Lake Pyroclastics
	Blackburn 1979b,	p. 29	Uphill Lake Pyroclastics and Flows
Sunshine Lake formation	Blackburn 1976,	p. 35 – 37	Mafic Sills
	Blackburn 1979b,	p. 33 – 38	Sunshine Lake Trachybasalts
Uphill Lake formation	Blackburn 1976,	p. 22 – 23	Cane Lake Pyroclastics
	Blackburn 1976,	p. 29 – 35	Supravolcanic Clastic Metasediments: Manitou Straits Metasediments
	Blackburn 1979b,	p. 39 – 44	Mosher Bay Metasediments
Mosher Bay formation	Blackburn 1976,	p. 29 – 35	Supravolcanic Clastic Metasediments: Manitou Straits Metasediments
	Blackburn 1979b,	p. 39 – 44	Mosher Bay Metasediments
Stormy Lake group	Blackburn 1979b,	p. 29 – 30	Washeibemaga Lake Pyroclastics and Autoclastics
	Blackburn 1979b,	p. 44 – 47	Washeibemaga Lake Metasediments
Boyer Lake group	Blackburn 1979b,	p. 26 – 27	Walmsley Lake Basalts
	Blackburn 1979b,	p. 30 – 31	Walmsley Creek Flows and Breccias
Blanchard Lake group	Blackburn 1976,	p. 12	Blanchard Lake Basalts
	Blackburn 1979a,	p. 11 – 12	Rector Lake Basalts
Upper Manitou Lake group	Blackburn 1976,	p. 19 – 20	Olsen Bay, Merrill Lake, and Early Lake Pyroclastics
	Blackburn 1976,	p. 20 – 21	Troutlet Lake Pyroclastics
	Blackburn 1976,	p. 12	Troutlet Lake Basalts
	Blackburn 1979a,	p. 19 – 26	Upper Manitou Lake Pyroclastics
	Blackburn 1979a,	p. 12 – 14	Upper Manitou Lake and Manitou Straits Basalts
	Blackburn 1979a,	p. 32 – 34	Metasediments
	Blackburn 1979b,	p. 31 – 32	Trafalgar Bay Pyroclastics
	Blackburn 1979b,	p. 27 – 28	Trafalgar Bay Basalts
Pincher Lake group	Blackburn 1976,	p. 10 – 11	Merrill Lake Basalts
(formations on the	Blackburn 1976,	p. 19 – 20	Olsen Bay, Merrill Lake and Early Lake Pyroclastics
NW limb of the	Blackburn 1979a,	p. 16 – 17	Johnar-Crooked Lake Basalts
Manitou Anticline)	Blackburn 1979a,	p. 26 – 28	Manitumeig Lake-Noonan Lake-Garnet Bay Pyroclastics
Benson Bay sub-group	Blackburn 1979a,	p. 15 – 16	Benson Bay Basalts
	Blackburn 1979b,	p. 27	Peekaboo Lake Basalts
	Blackburn 1979b,	p. 33	Peekaboo Lake Pyroclastics
	Blackburn 1979b,	p. 119 – 128	General Geology of the Gold Rock Camp
Beaverhead Island formation	Blackburn 1976,	p. 13 – 16	Beaverhead Island Basalts

Starshine Lake Sub-Group

Mafic metavolcanics underlie a large area that crosses the map-area in an arcuate form from the vicinity of Ponto Lake in the southwest, through Beck, Scattergood, Starshine, and Secret Lakes, to the vicinity of Thundercloud Lake in the east. The sub-group extends undetermined distances southwest and east of the synoptic map-area. It is a mixed assemblage of fine- to medium-grained massive and pillowed basalts, with amygdaloidal phases, and a number of porphyritic, brecciated, and coarse grained flow units, some of which can be traced continuously, allowing for fault offset, for up to 10 km along strike. Taking into account shallow northwesterly dips close to the contact with the Irene-Eltrut Lakes Batholithic Complex, and the steepening of dips in toward the centre of the metavolcanic-metasedimentary belt, the sequence attains a maximum thickness of 6700 m, measured across strike in a northwesterly direction from a point close to the southeast end of Scattergood Lake, to the top of the sequence near the southern shore of Sunshine Lake.

Emplacement of the Taylor Lake Stock has removed most of the upper part of the basaltic sequence in the east-central portion of the map-area. Movement along a number of northeast-trending faults, and one northwest-trending fault that has since been obliterated by emplacement of the Scattergood Lake Stock, has dissected the sequence into a number of offset segments.

The base of the sequence has apparently been removed by emplacement of granitic rocks of the Meggisi Lake lobe: portions of this base of the basaltic sequence occur as amphibolitic screens and xenolithic bodies in the granitic rocks at Meggisi Lake. The lowermost basaltic rocks, at all points along the margin with the granites, are strongly flattened pillowed metavolcanics, with conformable lenses and stringers of epidote. The flattened pillows dip away from the granite contact at low angles, maintaining angles of between 20 and 35 degrees for distances up to 1500 m from the contact, particularly in the vicinity of Secret Lake and south of Scattergood Lake. The resulting topography is one of steep-sided scarp-slopes facing toward, and shallow sloping dip-slopes facing away from, the granitic batholith. Individual flows cannot be made out in this sequence, which constitutes formation 1, but alternating sequences of massive and pillowed sections suggest that individual flows are composed of massive centres and pillowed tops, and are on the order of tens of metres thick. At many places, particularly east of Dorothy Lake, pillows cap the scarps, so that massive basalt is exposed in the scarp face, and only a narrow cap of pillows is found at the top.

Amygdaloidal phases of the pillowed flows were recognized to first occur about 600 m from the base of the sequence, both west and east of the Taylor Lake Fault. These rocks constitute formation 2 which is at a maximum 600 m thick, and absent in the southwest and extreme east.

Over most of their strike length, the amygdaloidal basalts are overlain by a sequence of non-amygdaloidal pillowed basalts, about 1000 m thick, but in the east a distinct coarse grained basalt unit, member 3A, about 200 m thick, intervenes, while west of the Taylor Lake Fault a similar, more discontinuous coarse-grained unit, member 3B, lies 200 m above the amygdaloidal basalts. Minor amounts of amygdaloidal basalt also occur in the sequence south of Taylor Lake. All of the above constitute formation 3. However, in the southwest, because of absence of formation 2, formations 1 and 3, being lithologically identical, merge.

A distinct, mappable massive to pillowed porphyritic basalt flow, formation 4, serves as an excellent stratigraphic marker. It can be traced discontinuously for over 20 km across the map-area from west of Scattergood Lake to east of Thundercloud Lake. The unit varies between 200 and 400 m in thickness, and is predominantly massive, with pillowed phases, but is porphyritic throughout, the phenocrysts being saussuritized plagioclase. The unit is intruded by the Scattergood Lake, Taylor Lake, and Thundercloud Lake Stocks, and is offset by the Taylor Lake Fault. Intrusion and presumably assimilation by the Taylor Lake Stock, and offset on the fault, accounts for a portion of the formation missing over a distance of 5 km in the east-central portion of the map-area. About 100 m above the porphyritic unit is a discontinuous 100 m thick zone of brecciation, part of which contains pillows and pillow fragments. This breccia unit has only been traced west of the Taylor Lake Fault to a point about 1 km west of Scattergood Lake. It is assigned here to formation 4.

Above the porphyry and breccia units is a 2500 m sequence of pillowed basalts, formation 5, most of which is non-amygdaloidal. It includes at least one discontinuous coarse basalt flow unit, member 5A, southeast of Starshine Lake. In the centre and west the pillowed basalts are overlain by a second distinct, mappable pillowed porphyritic basalt flow, formation 6, that can be traced continuously from just west of the Taylor Lake Fault, to the north end of Scattergood Lake, and its faulted equivalent lies 2,000 m to the south on the west side of the Scattergood Lake Stock, where it has been traced discontinuously to Beck Lake. The unit is on the order of 300 m thick, and similar in lithology to the porphyritic unit lower in the sequence.

Again in the centre and west, above the porphyritic basalts lies a mixed sequence, formation 7, beginning at base with pillowed basalts, but over a short distance passing upward into massive, non-pillowed basalt, and above this into coarser, gabbroic phases, much of which may be intrusive. Distinction in mapping between flow and intrusive phases was made rather arbitrarily on the basis of texture, and on presence of pillows in some places. In the southwest, narrow discontinuous lenses of intermediate to felsic pyroclastic rock, and minor interflow mudstones and sandstones occur toward the top of formation 7.

Etta Lake Formation

The base of the Etta Lake formation is here defined as the first occurrence, upward in sequence above Starshine Lake sub-group basalts, of continuous beds and members of clastic sedimentary rocks or pyroclastic rocks. The formation is a heterogeneous assemblage of such rocks interbedded with coarse mafic flows, some typically plagioclase-phyric and glomeroporphyritic, but all apparently devoid of pillows. At the top of the formation at its southwest end in the vicinity of Meridian Bay is a 15 to 30 m thick chert member that can be traced discontinuously for more than 5 km.

Faulting has severely disrupted the formation, making correlation of individual members a difficult task. The formation is terminated in the east by the Scattergood Lake Stock, but continues southwestward an undetermined distance out of the map-area. "Volcanogenic sediments" were mapped by McWilliams and Ali (*in* Sage *et al.* 1974) as a southwestward continuation of the formation at least another 3 km.

During field checking in 1976 at Glass Bay the author observed sedimentary features, such as graded bedding, ripple marks, and load casts, in units previously (Blackburn 1976) mapped as pyroclastic. It is therefore probable that most, if not all, of the clastic material in the formation is at least reworked rather than primary pyroclastic debris.

Glass Bay Formation

The base of the Glass Bay formation is defined as the first basalts that lie above clastic rock of the Etta Lake formation. Clastic material is therefore absent in this formation. The formation is composed of pillowed lavas, and associated aquagene breccia, and is on the order of 1,200 m thick. Some of the breccia occurs as more or less continuous members parallel to strike.

MANITOU GROUP

This group has been the subject of a sedimentological study by Teal (Teal and Walker 1977; Teal 1979). Teal followed the lead of Thomson (1933) in delineating the extent of the group, though confining the name to the western portion of Thomson's Manitou "Series". Teal and Walker (1977) noted that the underlying volcanic rocks (Wapageisi Lake group of this report) "form a base upon which the group rests". However, Teal (1979) did not map the base of the group east of the Scattergood Lake Stock, although he recognized that pyroclastic rocks on the south shore of Uphill Lake, not included by Thomson in his Manitou "Series", are similar to rocks at Cane Lake that were included by Thomson in his Manitou "Series", and therefore should be included in the Manitou group. Mapping by the author (Blackburn 1979b) has clearly delineated these pyroclastics and intrusion into them, and therefore the base of the group.

The Manitou group has been described by Teal and Walker (1977) to comprise four units. These have been proposed as formations[1] by Teal (1979), and Teal's terminology has been mostly followed here. Although the boundaries of the formations are similar to those of Teal, the lithologic interpretations are not in all cases the same. Also, because Teal did not investigate lithologies on the south sides of Uphill and Sunshine Lakes he did not delimit the extent of the lowermost of his formations, the Cane Lake formation. A second unit, the Sunshine Lake formation, is considered here to have subvolcanic equivalents in the form of sills at Sunshine Lake and Cane Lake. Only the subaerial rocks were investigated by Teal. The felsic Sunshine Lake intrusion is here also considered to be part of the Manitou group, because of its subvolcanic relationship to felsic volcanic rocks within the Cane Lake formation. Teal and Walker (1977) and Teal (1979) have further delineated two members within their unit 3, the Uphill Lake formation. Insufficient study was carried out by the present worker to distinguish these members. A fourth unit, the Mosher Bay formation, overlies the Uphill Lake formation. In addition to descriptions of lithologies that comprise these formations made previously by this author (Blackburn 1976, 1979b), the reader is referred to the more detailed descriptions and discussions of sections by Teal (1979).

Cane Lake Formation

The Cane Lake formation is predominantly composed of heterolithologic, intermediate, pyroclastic rocks, though minor felsic flows, and epiclastic sedimentary rocks are also included within its confines. It is intruded by mafic and felsic subvolcanic rocks that are not considered to be part of the formation.

The base of the formation is defined where clastic rocks overlie the thick basaltic flow sequence of the Glass Bay formation. This base is disrupted by intrusion of the Scattergood Lake Stock. Southwest of the stock the contact appears to be sharp, and where observed by Blackburn (1976, Photo 10) appeared to be erosional. East of the stock the base is either cut out by faulting, or by intrusion of mafic and felsic subvolcanic bodies.

Sunshine Lake Subvolcanic Intrusion

The Sunshine Lake intrusion, described under the heading "Sunshine Lake Porphyry" in the report on the Boyer Lake-Meggisi Lake area (Blackburn 1979b) is considered to be the subvolcanic equivalent of massive felsic volcanic rocks at the west end of Sunshine Lake, and thus within the Cane Lake formation.

---

[1] Teal and Walker (1977) have stated that formal definitions of formations within the Manitou group will be forthcoming. Although Teal (1979) has satisfied the requirements of the code of stratigraphic nomenclature (ACSN 1961) in so far as descriptions of these formations are concerned, because the descriptions have not been published as of date of writing (December 1979) in conformity with Article 13, remark c (ACSN 1961, p.653) the formations must be regarded as informal.

Sunshine Lake Formation and Associated Sills

The Sunshine Lake formation is a distinctive mafic flow unit that, although in places quite narrow (less than 30 m), can be traced for over 15 km along strike. Its sharp base conveniently defines the top of the underlying Cane Lake formation, though because it terminates laterally in a southwesterly direction near Watson's Falls at Cane Lake, the upper boundary of the Cane Lake formation is transitional into the overlying Uphill Lake formation beyond this point.

During 1972 field mapping, the author (Blackburn 1976) classified all of this unit southwest of Rush Bay of Uphill Lake as a mafic sill. Subsequent discussion with P.R. Teal and mapping in 1975 (Blackburn 1979b) of the eastern end of the unit led to the interpretation that the unit is entirely extrusive. However, another mafic unit that outcrops on Cane Lake and that was mapped as a sill in 1972, retains that status, and is further discussed below.

The mafic sill at Cane Lake is here newly correlated with an ultramafic to lamprophyric sill at the south shore of Sunshine Lake, discussed as the northerly of two sills under the heading "Sunshine Lake Gabbros, Peridotites, and Lamprophyres" in Blackburn (1979b).

It is noted here that some of the ultramafic rocks may be extrusive, since breccias with ultramafic clasts are found in association with them. Relationships between the ultramafic rocks and the lamprophyres were not fully established in 1975 (Blackburn 1979b) and no further observations were made during the 1976 field season. Also noted in 1975 was the fact that the lamprophyres are in places dike-like, and in contact with the Sunshine Lake felsic porphyry, therefore post-dating it.

Uphill Lake Formation

The Uphill Lake formation is composed of volcanic-clast and polymictic conglomerates, pyroclastic rocks, and sandstone and argillite. The work of Teal and Walker (1977) and Teal (1979) has formed the basis for formational terminology used here, and has necessitated subdivision and re-arrangement of the lithologies discussed in the earlier (Blackburn 1976, 1979b) reports of the present author.

The base of the Uphill Lake formation differs structurally and lithologically along its length. North of Sunshine Lake it is gradational upward from the Sunshine Lake trachybasalt formation, where rubbly volcanic breccia passes upward into volcanic-clast conglomerate. North of Uphill Lake, the lower contact is sharp, with sandstone, mudstone, or pyroclastic rocks variably occurring at base, on top of the Sunshine Lake formation. West of Cane Lake the Sunshine Lake formation terminates, and the present author can find no good criteria upon which to distinguish the Uphill Lake formation from the Cane Lake formation: Teal and Walker (1977) and Teal (1979) came to the same conclusion.

Teal and Walker (1977) and Teal (1979) have been able to recognise three facies of depositional environment within the sedimentary rocks of the Uphill Lake formation: a braided fluvial environment in the east, to the north of Sunshine Lake; a lacustrine facies near the base of the formation in the central portion, north of Uphill Lake; and an alluvial fan environment constituting the bulk of the formation in the central portion, to the north of Uphill and Cane Lakes. The alluvial fan environment passes southwestward into the pyroclastic part of the formation. At the top of the formation they have distinguished (Teal 1979, p.vi) "a highly variable group (member) of siltstones, argillites, volcanogenic and quartzose sandstones, conglomerates, tuffs, and tuff-breccias ..... From its stratigraphic position, the member could be interpreted as a coastal or shallow marine deposit, but, unfortunately it is entirely without any diagnostic features".

Mosher Bay Formation

Teal and Walker (1977) and Teal (1979) have defined this formation as "a group of argillites, sandstones and conglomerates of the Resedimented Association ... at the top of the Manitou Group". The present author concurs with this definition; however, because insufficient study was carried out by the present author to define the base of the formation on this basis, the position of this contact is taken mostly from the work of Teal (1979). Certain observations are made here that bear on the boundary as defined by Teal. Firstly, it is noticeable that the transition upward in the conglomerates of the Manitou group from volcanic-clast to polymictic is fairly abrupt along most of the strike length, and corresponds to the Uphill Lake-Mosher Bay formational boundary. This conforms with the observation that the resedimented conglomerates, which occur interbedded with graded sandstones of turbidite type, are markedly polymictic: the combination of resedimented conglomerates and graded sandstones are characteristic of the "Resedimented Association" (Turner and Walker 1973; Walker 1976). However, in the east, polymictic conglomerates, mapped by Blackburn (1979b) occur lower in the sequence, well below the formation boundary as delimited by Teal (1979), and within the Uphill Lake formation. Graded sandstones of turbidite type are not associated with these conglomerates. Teal and Walker (1977) and Teal (1979) have interpreted this east end of the Uphill Lake formation to be of braided fluvial sedimentary facies association.

The second observation that bears on the boundary is that numerous faults mapped by Blackburn (1979b) but unrecorded by Teal (1979), offset the boundary. In general this offset is probably minor, however.

STORMY LAKE GROUP

The Stormy Lake group is named here after a large lake east of the present map-area. Reconnaissance mapping by Thomson (1933) showed that clastic lithologies at Stormy Lake are continuous with those discussed here under this heading. Thomson referred them to his Manitou "Series", but, as discussed above, a new group name is considered necessary.

At the present state of detailed mapping, there is insufficient coverage eastward along strike to subdivide the very distinctive and differing lithologies within the group into formations.

Following reconnaissance mapping by Thomson (1933), Bertholf (1946) divided the clastic sequence at Washeibemaga Lake into two units: the lower, Sagenak conglomerate (after Sagenak Lake, now Seggemak Lake) and the upper, Dark Horse conglomerate (after an informally named lake east of Washeibemaga Lake). Within the lower unit he included a large variety of rocks, including a very distinctive accumulation of quartz-feldspar porphyry clasts in a similar matrix. Thomson (1933) had included this accumulation within a quartz porphyry stock, not recognising its subtle but distinctive difference. Bertholf (1946) interpreted this accumulation to be an arkosic breccia derived from the underlying porphyry stock, the contact between the two being a "graded unconformity". Subsequent mapping by the author (Blackburn 1979b) and McMaster (1978) led to a further re-interpretation of this accumulation as a felsic autoclastic breccia, with the current author favouring a pyroclastic origin for much of the upper part of it. Vague, but definite, layering was mapped by the author in this accumulation, a feature that is difficult to reconcile with an entirely autoclastic origin.

The author (Blackburn 1979b) interpreted the upper half of the group within the map-area to be almost entirely epiclastic. McMaster (1978), however, interpreted almost all of these same rocks to be pyroclastic. Resolution of these discrepancies and differences awaits further detailed mapping east of the present map-area.

Thundercloud Lake Subvolcanic Intrusion

The Thundercloud Lake intrusion, described under "Thundercloud Lake Porphyry" in the report on the Boyer Lake-Meggisi Lake area (Blackburn 1979b), is the underlying, subvolcanic phase of the felsic autobreccia described above. It is therefore an integral part of the Stormy Lake group. McMaster (1978) has also described the porphyry in some detail.

BOYER LAKE GROUP

Reconnaissance mapping by Thomson (1933) and subsequent detailed and reconnaissance mapping by the present author (Blackburn 1979b; Trowell *et al.* 1980) has delineated a distinctive group of predominantly mafic metavolcanics, intruded by mafic to ultramafic plutonic rocks, that appears to be bounded by the Manitou Straits Fault in the west, the Manitou and Stormy Lake groups in the south, and pyroclastic rocks at Kawashegamuk Lake to the east of the map-area. The northern limit is unclear at the present stage of detailed mapping. The name Boyer Lake group is proposed here (recognising that its limits are not entirely known at present), after a large lake in the northeast part of the present map-area.

Although distinct volcanic units are recognised within the present map-area, no formations are established here, pending further detailed mapping to the east, currently being undertaken by the author (Blackburn 1979c).

Gabbro Lake, Mountdew Lake, Mosher Bay, and Kenny Lake Gabbros

Lithologies of the gabbro bodies southeast of the Manitou Straits Fault were described under the general heading of "Metamorphosed Mafic and Ultramafic Intrusive Rocks" in the report on the Boyer-Meggisi Lakes area (Blackburn 1979b). The Gabbro Lake body was there referred to under the heading "Boyer Lake Gabbros"; this name is considered inappropriate if the name Boyer Lake is to be applied as the name of the group they intrude. McMaster (1975) referred to the body as the Gabbro Lake Sill.

These bodies, exemplified by the Gabbro Lake and Mountdew Lake Gabbros, are clearly not subvolcanic equivalents of the enclosing mafic metavolcanics. McMaster (1975) has shown the Gabbro Lake body to be a fractionally crystallized and differentiated, horizontally emplaced sill that has subsequently been tilted into its present approximately vertical position. However, emplacement of both this sill, and the Mountdew Lake Gabbro, post-dates the only demonstrable episode of folding in the metavolcanics, since they both cross-cut the Kamanatogama Syncline (discussed below). Since at least the Gabbro Lake body was emplaced horizontally, it follows that the Kamanatogama Syncline has itself been tectonically disturbed, as discussed later in this report under "Geologic Evolution of the Manitou Lakes Area".

Northwest of the Manitou Straits Fault

Supracrustal and subvolcanic rocks northwest of the Manitou Straits Fault are here assigned to three lithostratigraphic groups. Abundant top determinations in metavolcanics and tracing of contacts show that the three groups are folded about a northeast trending and plunging anticline, such that the oldest occurs in the core of the fold. They are discussed below in ascending order of succession.

BLANCHARD LAKE GROUP

The Blanchard Lake group is named after a lake that occurs in the middle of the group. At the present stage of mapping, formations have not been delineated within the group, though it is apparent that a variety of mafic flow lithologies, many of which have been traced along strike on the order of three or more kilometres, are present. The base of the sequence is not known, since it lies in the core of the anticline. The sequence as exposed is at least 1200 m thick.

The group is characteristically mafic. A 300 m thick coarse-grained mafic unit within the sequence on the southeast limb of the fold that was named the Doyle Lake Sill in an earlier report (Blackburn 1976) was not encountered along strike to the northeast (Blackburn 1979a). Neither was it found across the fold structure on the northwest limb of the anticline.

UPPER MANITOU LAKE GROUP

The Upper Manitou Lake group is named after the large lake which occupies much of the fold-nose of the Manitou Anticline. The name should not be confused with the Manitou group, described previously.

The group is composed of a distinctly heterogeneous assemblage of mafic to felsic metavolcanics and minor metasediments. Because of both lack of continuous outcrop, and intense shearing, the definition of formations within the group is at present conjectural. The group name is proposed with the conviction that with further more detailed study within the confines of the group, several formations could be delimited.

The base of the group is defined at the distinct hiatus where Blanchard Lake mafic metavolcanics pass upward into intermediate pyroclastics. Metasediments define this transition in at least one locality, on the west shore of Reliance Bay of Upper Manitou Lake (Blackburn 1979a), and possibly also at Doyle and Shaughnessy Bays of Lower Manitou Lake, where an "argillitic or fine tuffaceous unit" and "volcanic clastic sandstones and siltstones" occur "at the base of the pyroclastic sequence" (Blackburn 1976, p.20).

The group is considered to include predominantly pyroclastic material, with less continuous mafic flow members or formations. As such its upper boundary is therefore defined where mafic flows of the overlying Pincher Lake group commence. Much of this boundary, in the west and north, is marked by mafic metavolcanics of the overlying Stony Island formation. However, along the southeast side the group boundary is somewhat conjectural, because of intense shearing associated with the Manitou Straits Fault, and the possibility of faulting out or repetition of sequences. The boundary is drawn along the northwest side of Manitou Straits, east of Upper Manitou Lake, and more tentatively along the same, northwest, side of the Straits in the vicinity of Beaverhead Island, at the base of the overlying mafic metavolcanics of the Beaverhead Island formation.

Frenchman Island Subvolcanic Intrusion

Also considered to be an integral part of the Upper Manitou Lake group is a body of microgranodiorite and micro quartz diorite porphyry that underlies most of Frenchman Island of Upper Manitou Lake, and some of the surrounding islands. Lithologies that comprise this body are discussed under "Frenchman Island Subvolcanic Core" in the report on the Upper Manitou Lake area (Blackburn 1979a). As discussed there (Blackburn 1979a, p.31), "the Frenchman Island body was probably emplaced in a feeder or vent from which much of the pyroclastic material at Upper Manitou Lake was ejected".

PINCHER LAKE GROUP

The Pincher Lake group is named here after a lake in the north-central part of the present map-area. The extent of northern continuation of the group is not known because of lack of detailed mapping, but reconnaissance work by the present author in conjunction with a regional stratigraphic study (*in* Trowell *et al.* 1980) suggests that the group may be continuous or interdigitate with a thick sequence of volcanic rocks (termed "Wabigoon Volcanics" in Trowell *et al.* 1980) extending north and west to Wabigoon Lake (Satterly 1941) and Eagle Lake (Moorhouse 1939).

The present level of mapping allows the Pincher Lake group to be divided into seven formations, and a sub-group. It is also intruded by a gabbro body. These lithostratigraphic units are combined into a single group because the six formations on the northwest limb of the Manitou Anticline appear to represent a cyclical repetition, and the sub-group, which may be divisible into formations, interdigitates with these formations. Also the Pincher Lake group is lithologically distinctly different from the underlying Upper Manitou Lake group.

Without giving valid reasons, Goodwin (1965) interpreted a northeast-trending syncline to lie within the area encompassed by the six formations on the northwest limb of the Manitou Anticline. During the detailed field mapping by the present author (Blackburn 1979a), top determinations were confined to near Upper Manitou Lake, and therefore did not provide further information on the presence of such a fold. In the absence of such corroborating evidence, and because of abundant evidence from pillow facings that the Benson Bay sub-group, with which the six formations appear to interdigitate, faces northwest, it is concluded here that the sequence of six formations homoclinally faces northwest also.

Stony Island Formation

The Stony Island formation is a basalt flow sequence, on the order of 1000 m thick, with minor pyroclastic lenses. Increasing grade of metamorphism along strike within the formation toward the southwest obscures observations, but it appears that pillowed lavas, although characteristic in the northeast in the vicinity of Benson Bay, are rare to absent in the southwest, in the vicinity of Merrill Lake. Plagioclase-phyric flows are characteristic, and at least one such unit, on the order of 100 m wide, can be followed fairly continuously along the southeast side of Merrill Lake, a distance of almost 4 km.

Northeastward the formation passes into the Benson Bay sub-group where it has not been delineated, probably because the overlying Garnet Bay formation pinches out in this vicinity. Similarly in the southwest the overlying formation terminates east of Kaminni Lake, where Stony Island formation basalts merge with similar lithologies of the Dog Lake formation.

Garnet Bay Formation

A distinct and continuous unit of intermediate tuff-breccia conformably overlies the Stony Island formation over a strike length of about 11 km, from east of Kaminni Lake in the southwest, via Merrill Lake, to Garnet Bay of Upper Manitou Lake in the northeast. At Garnet Bay these coarse pyroclastics pass laterally into lapilli-tuffs and tuffs, becoming more felsic northeastward. A thin felsic flow or pyroclastic member extends to Benson Bay.

Dog Lake Formation

The Dog Lake formation is in conformable contact with the underlying Garnet Bay formation. It consists of mafic metavolcanics, some of which are plagioclase-phyric. It extends from Garnet Bay in the northeast to Dog Lake in the southwest. Amphibolite facies metamorphism, coupled with deformation, has obscured relationships in the southwest, but the formation appears to merge with the Johnar Lake formation, of similar mafic flow lithology, in this area.

Noonan Lake Formation

A distinct and continuous unit of intermediate tuff, lapilli-tuff, and tuff-breccia occupies the southeast shores of Noonan and Johnar Lakes. It lies conformably above the underlying Dog Lake formation. Similar pyroclastics, and minor flows, some felsic, extend from Johnar to Harper Lake, and minor amounts of mafic flows are intercalated. These units are combined under the Noonan Lake formation.

Johnar Lake Formation

The Johnar Lake formation comprises mafic metavolcanics that conformably overlie the Noonan Lake formation, and includes minor pyroclastic lenses. It is terminated to the northeast at Marius Lake by diorite, and by gabbro of the Mitchell Lake intrusion. To the southwest it merges with similar lithologies of the Dog Lake formation. Midway up the sequence, mapping by the author (Blackburn 1979a) indicates a passage upward into amygdaloidal lavas.

Manitumeig Lake Formation

A gradational boundary exists between underlying Johnar Lake formation mafic flows and the largely pyroclastic Manitumeig Lake formation, with minor tongues of mafic metavolcanics. Also present, immediately adjacent to the invading Atikwa Batholith, are two narrow felsic flow members.

Strongly deformed and metamorphosed pyroclastic and mafic flows that occur as a xenolithic mass northwest of Big Manitumeig Lake, and described under "Big Manitumeig Lake Pyroclastics" in the report on the Upper Manitou Lake area (Blackburn 1979a) are probably the detached continuation, at least in part, of the Manitumeig Lake formation. In addition, pyroclastics that occur at the margin of the metavolcanic belt east of Kekekwa Lake may also be its lateral equivalent, although grouped here under the Benson Bay sub-group.

Benson Bay Sub-Group

Northeastward of the six formations discussed previously is a mixed sequence of flows and pyroclastics that wraps around the Manitou Anticline and extends southwestward along the Manitou Straits. It is bordered on its southeast side by the Manitou Straits Fault.

At about the vicinity of Watson's Narrows on Manitou Straits the sub-group can be considered to terminate. From this point southwestward the Pincher Lake group is divisible into a formation (Beaverhead Island formation), and an unassigned sequence of predominantly pyroclastic rocks, with minor epiclastic sedimentary rocks, and mafic flows, previously described under "Beaverhead Island Pyroclastics" in the report on the Lower Manitou-Uphill Lakes area (Blackburn 1976).

Beaverhead Island Formation

The Beaverhead Island unit is singled out as a formation because of its mappable continuity as a sequence of mafic flows over a strike length in excess of 10 km. Lithologies are discussed under "Beaverhead Island Basalts" in the report on the Lower Manitou-Uphill Lakes area (Blackburn 1976).

Mitchell Lake Gabbro

Lithologies of the Mitchell Lake gabbro body are described in the report on the Upper Manitou Lake area (Blackburn 1979a). Relative age of emplacement of the gabbro was not ascertained. The body is clearly intrusive into the Pincher Lake group, but whether it represents a subvolcanic phase is uncertain.

PETROCHEMISTRY

Petrochemical work by the author has been confined to the volcano-sedimentary or supracrustal belt, and to felsic and intermediate subvolcanic intrusions, one of which was analysed as part of a thesis by McMaster (1978). Also, one of the gabbroic intrusions was analysed as part of another thesis of McMaster (1975). Granitic rocks in the present map-area were chemically analysed as parts of theses by Pichette (1976), Birk (1978), and Sabag (1979).

Volcano-Sedimentary Belt

Chemical analyses were made of selected samples during detailed geological surveys (Blackburn 1976, 1979a, b). Use is made of these here to augment results from a new set of analyses. These new analyses are of 71 samples (Table 4) taken from four sections across various parts of the belt during the 1976 field season. The samples were collected, where possible, every quarter mile (400 m) across strike. Locations of samples are shown on Figure 3 (Chart A, back pocket). The intention was to obtain data on chemical change with stratigraphic height, at regular intervals, within each section. Sections were only run in predominantly mafic flow sequences, and not in predominantly pyroclastic sequences, because of the inherent difficulties in sampling pyroclastic rocks due to grain size and heterogeneity, and due to lateral facies variation. Numerous samples had been previously taken from one such predominantly pyroclastic sequence, at Upper Manitou Lake (Blackburn 1979a); no consistent trends can be established from this data, and a wide scatter of results is evident (see Figure 4e, f).

All samples taken by the author to date from within the map-area, with the exception of those from the Sunshine Lake formation, are presented on cation plots (after Jensen 1976), and AFM diagrams (after Irvine and Baragar 1971). Samples collected prior to the synoptic study are plotted without sample numbers (Figure 4) except where required to clarify the text: chemical analyses are tabulated and plotted in the detailed reports (Blackburn 1976, 1979a, b). A suite of samples from a gabbroic body analysed by McMaster (1975) is also presented (Figure 4c, d) for comparison. For the four new sections, all samples plotted on the diagrams (Figure 5) are numbered as in Table 4. For completeness, all samples taken from the Sunshine Lake formation by the present author and by Teal (1979) are presented on diagrams (Figure 7) that classify them according to the method of Irvine and Baragar (1971).

Samples taken from the four new sections are also plotted on diagrams (Figure 6) that show variations in oxide weight percents, differentiation index, and crystallization index, against stratigraphic height.

WAPAGEISI LAKE GROUP

AFM and cation plots (Figures 4a, b, 5a, b, c, d) for samples from the Wapageisi Lake group show clearly their tholeiitic chemistry. The cation plots (based on Jensen 1976) also

TABLE 4. CHEMICAL ANALYSES (WEIGHT PERCENT) OF METAVOLCANICS FROM THE BECK, STARSHINE, DOYLE, AND JONAS LAKE SECTIONS OF THE MANITOU LAKES AREA. ANALYSES BY GEOSCIENCE LABORATORIES, ONTARIO GEOLOGICAL SURVEY. FOR SAMPLE LOCATIONS SEE FIGURE 3.

Major Elements	B1	B2	B3	B4	B5	B6	B7	B8	B9	B10	B11	B12	B13	B14	B15	B16	B17	B18	B19	B20
SiO_2	48.0	56.3	47.5	49.3	49.8	50.8	49.2	50.7	49.1	48.9	49.3	49.4	48.7	47.6	47.9	49.9	48.5	49.2	48.4	48.3
Al_2O_3	15.3	14.6	12.2	11.7	14.3	14.2	13.4	13.3	14.3	15.5	16.2	15.8	15.4	17.0	15.8	15.1	15.5	14.8	15.3	15.2
Fe_2O_3	11.6	9.12	12.8	12.3	12.6	13.5	12.2	10.3	13.1	11.5	12.1	12.4	12.4	12.7	12.3	15.2	12.1	12.9	13.6	12.2
MgO	7.07	6.87	13.3	12.0	7.81	6.79	10.8	6.96	7.87	7.47	3.78	7.39	8.57	7.22	9.10	5.55	6.61	6.42	7.51	7.01
CaO	14.3	8.89	10.8	10.9	11.0	10.5	11.6	16.6	10.0	13.7	15.3	12.3	12.0	10.3	10.2	8.69	13.2	11.4	11.6	13.0
Na_2O	0.99	2.60	1.34	1.99	2.79	1.96	1.66	0.37	2.51	1.19	0.05	2.09	1.63	1.46	1.49	1.49	1.48	2.41	1.35	1.52
K_2O	0.36	0.49	0.17	0.09	0.25	0.17	0.08	0.03	0.51	0.20	0.02	0.10	0.08	0.07	0.07	0.13	0.11	0.15	0.05	0.08
TiO_2	0.74	0.50	0.58	0.57	0.71	0.93	0.61	0.57	0.96	0.78	0.77	0.68	0.65	0.77	0.65	1.27	0.82	0.89	0.89	0.73
P_2O_5	0.08	0.07	0.08	0.08	0.09	0.11	0.08	0.08	0.09	0.08	0.10	0.08	0.08	0.09	0.08	0.13	0.10	0.10	0.09	0.09
MnO	0.18	0.16	0.20	0.19	0.18	0.20	0.21	0.23	0.22	0.19	0.20	0.19	0.21	0.22	0.20	0.22	0.18	0.24	0.22	0.19
L.O.I.	1.91	0.87	1.31	0.95	0.63	1.03	0.95	2.07	1.39	1.39	2.23	0.87	1.07	3.11	2.63	3.15	1.71	1.71	1.99	2.55
Total	100.5	100.5	100.3	100.1	100.2	100.2	100.8	101.2	100.0	100.9	100.0	101.3	100.8	100.5	100.4	100.8	100.3	100.2	101.0	100.9

Major Elements	B21	B22	S1	S2	S3	S4	S5	S6	S7	S8	S9	S10	S11	S12	S13	S14	S15	S16	S17	D1
SiO_2	47.2	50.5	48.8	48.1	53.4	49.8	51.0	49.6	48.4	43.7	48.4	48.2	48.3	50.2	49.0	50.2	50.1	49.3	49.0	47.4
Al_2O_3	16.7	13.1	11.9	11.8	13.0	13.4	15.6	14.2	15.7	17.3	16.0	15.7	16.4	15.6	16.6	15.7	14.8	17.6	15.3	16.1
Fe_2O_3	12.8	16.7	12.6	12.7	12.3	11.8	10.3	12.0	11.8	14.4	11.7	12.7	12.9	12.5	13.4	12.1	12.4	11.8	12.4	12.9
MgO	7.11	4.98	13.0	12.9	7.06	9.77	8.36	9.73	8.96	9.72	7.86	9.35	6.66	7.04	5.38	6.41	6.07	6.14	7.18	7.76
CaO	11.1	7.35	10.8	10.7	9.77	11.4	11.7	11.1	11.4	8.82	12.8	9.37	12.3	11.0	11.6	10.8	12.7	11.2	13.2	10.2
Na_2O	1.79	3.03	1.80	1.71	2.76	2.25	2.49	1.83	1.26	2.11	1.51	2.64	1.75	2.08	1.72	1.91	1.90	1.77	1.37	1.81
K_2O	0.06	0.22	0.05	0.16	0.18	0.17	0.31	0.12	0.52	0.29	0.09	0.10	0.11	0.09	0.25	0.32	0.13	0.17	0.07	0.06
TiO_2	0.76	1.59	0.56	0.59	0.82	0.62	0.54	0.64	0.79	0.92	0.62	0.68	0.93	0.87	0.93	0.86	0.82	0.82	0.74	0.90
P_2O_5	0.09	0.14	0.08	0.08	0.11	0.08	0.08	0.08	0.09	0.09	0.08	0.08	0.10	0.10	0.10	0.10	0.10	0.09	0.09	0.10
MnO	0.20	0.25	0.20	0.22	0.24	0.19	0.18	0.23	0.18	0.22	0.19	0.22	0.21	0.21	0.26	0.20	0.20	0.18	0.21	0.20
L.O.I.	3.11	2.43	0.71	1.83	0.47	0.91	0.95	1.15	1.23	3.23	0.75	1.59	0.55	0.63	1.19	1.63	1.43	1.47	1.39	3.11
Total	100.9	100.3	100.5	100.8	100.1	100.4	101.5	100.7	100.3	100.8	100.0	100.6	100.2	100.3	100.4	100.2	100.6	100.5	100.9	100.5

Major Elements	D2	D3	D4	D5	D6	D7	D8	D9	D10	D11	D12	D13	D14	D15	J1	J2	J3	J4	J5	J6
SiO_2	48.3	49.4	47.0	46.3	46.5	49.2	52.4	51.0	50.1	52.1	52.4	49.6	47.9	49.0	52.7	64.1	68.6	48.6	50.3	47.5
Al_2O_3	15.6	15.2	14.9	14.0	14.5	15.3	12.9	14.1	15.5	13.6	14.3	15.0	15.6	15.2	15.5	16.2	15.3	12.2	13.7	15.2
Fe_2O_3	13.5	13.0	11.4	12.5	12.1	12.5	12.5	14.2	14.8	14.8	13.0	14.0	12.3	13.1	12.3	5.75	3.86	12.4	15.9	14.0
MgO	7.80	6.83	7.38	7.29	11.4	6.66	6.11	6.24	6.15	5.51	5.02	7.51	8.03	6.98	3.73	1.53	0.99	3.90	5.81	7.36
CaO	8.30	10.5	11.8	8.93	11.3	12.2	8.80	9.32	7.89	9.13	9.64	9.15	9.80	11.4	4.82	3.14	2.43	12.7	8.05	10.0
Na_2O	2.95	2.55	1.12	1.66	1.15	1.87	2.73	2.82	3.30	2.22	2.88	2.27	2.30	2.06	4.95	5.07	3.19	1.63	2.01	1.84
K_2O	0.14	0.18	0.04	0.13	0.03	0.10	0.06	0.20	0.14	0.18	0.09	0.14	0.14	0.17	0.04	0.77	2.38	0.05	0.17	0.34
TiO_2	1.18	1.12	0.81	1.07	0.36	1.07	1.34	1.40	1.62	1.58	1.22	1.29	0.97	1.09	1.19	0.73	0.38	1.10	1.44	0.91
P_2O_5	0.12	0.13	0.10	0.12	0.06	0.12	0.15	0.15	0.17	0.16	0.17	0.13	0.11	0.12	0.18	0.28	0.15	0.13	0.12	0.10
MnO	0.20	0.20	0.18	0.18	0.20	0.20	0.17	0.20	0.23	0.21	0.18	0.20	0.19	0.19	0.16	0.06	0.08	0.23	0.24	0.21
L.O.I.	2.99	2.07	5.95	8.06	3.15	1.55	3.59	1.31	1.35	0.55	0.83	1.31	3.23	1.19	5.43	2.71	3.39	7.54	2.91	2.75
Total	101.1	101.2	100.7	100.8	100.7	100.8	100.7	100.9	101.2	100.0	99.7	100.6	100.6	100.5	101.0	100.3	100.7	100.5	100.6	100.2

Major Elements	J7	J8	J9	J10	J12	J13	J14	J15	J16	J17	J18
SiO_2	55.4	44.4	47.7	48.3	48.1	52.2	58.3	55.5	48.3	54.3	51.0
Al_2O_3	15.1	17.3	15.4	14.8	15.3	16.3	16.5	14.5	16.4	15.8	15.5
Fe_2O_3	9.45	11.0	12.9	13.6	12.6	11.0	8.57	10.1	13.5	10.7	12.8
MgO	5.12	4.30	8.02	7.63	8.04	5.55	3.10	5.30	5.70	6.13	4.61
CaO	6.75	9.77	7.42	10.6	11.2	9.64	7.38	10.6	10.6	8.03	10.9
Na_2O	2.81	4.43	1.76	1.75	1.12	1.81	4.51	2.43	2.05	3.30	2.35
K_2O	0.13	0.34	0.29	0.08	0.05	0.16	0.26	0.20	0.72	0.26	0.25
TiO_2	1.03	1.62	0.90	0.96	0.82	0.81	1.13	1.01	1.10	1.09	1.23
P_2O_5	0.25	0.37	0.10	0.10	0.09	0.17	0.25	0.21	0.21	0.05	0.31
MnO	0.12	0.27	0.21	0.21	0.19	0.21	0.12	0.17	0.26	0.16	0.18
L.O.I.	5.07	6.95	6.27	2.47	2.59	2.43	0.51	0.71	1.35	0.75	0.59
Total	101.2	100.7	101.0	100.5	100.1	100.3	100.6	100.7	100.2	100.6	99.7

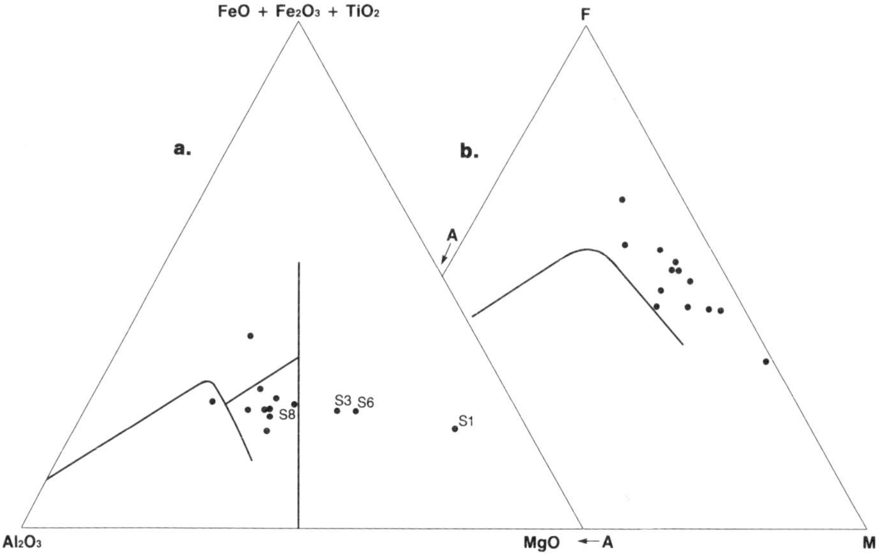

Wapageisi Lake group.

Figure 4—*Chemical classification of metavolcanics of the Wapageisi Lake, Boyer Lake, Upper Manitou Lake, and Pincher Lake groups. Cation plot after Jensen (1976); AFM plot after Irvine and Baragar (1971).* ***a,b.*** *Wapageisi Lake group (data from Blackburn 1976, 1979b). Numbered samples from Blackburn (1979b).*

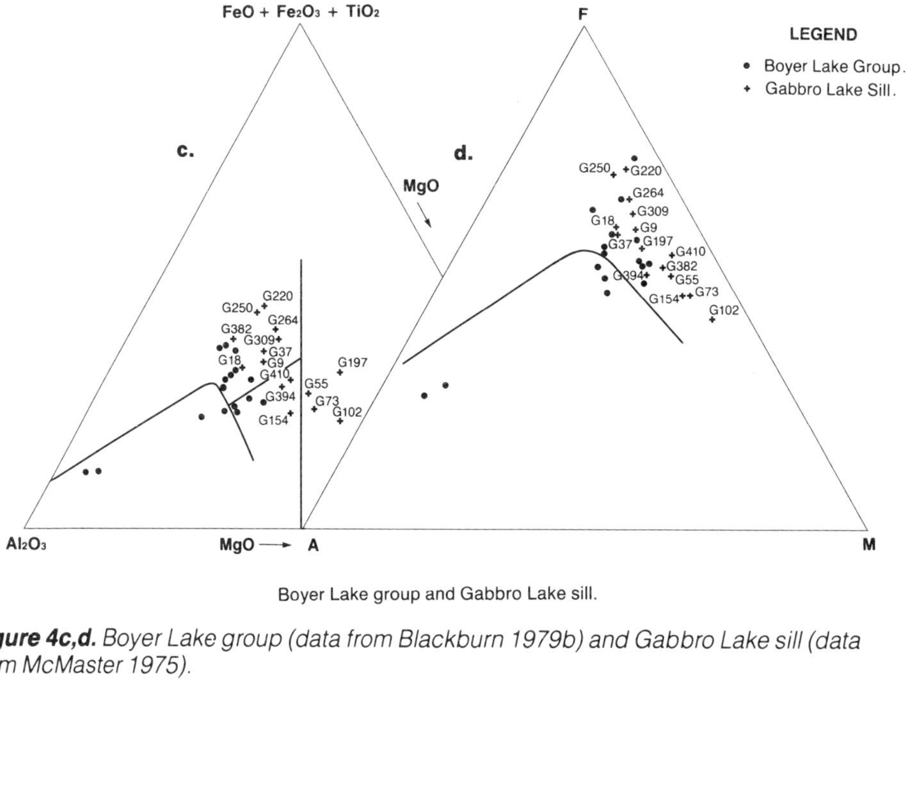

Boyer Lake group and Gabbro Lake sill.

Figure 4c,d. *Boyer Lake group (data from Blackburn 1979b) and Gabbro Lake sill (data from McMaster 1975).*

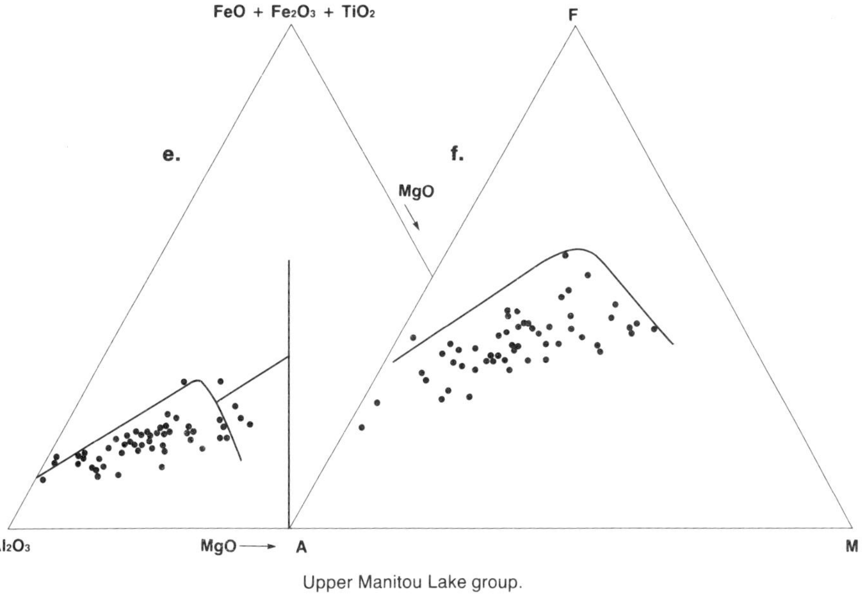

Figure 4e,f. *Upper Manitou Lake group (data from Blackburn 1976, 1979a,b).*

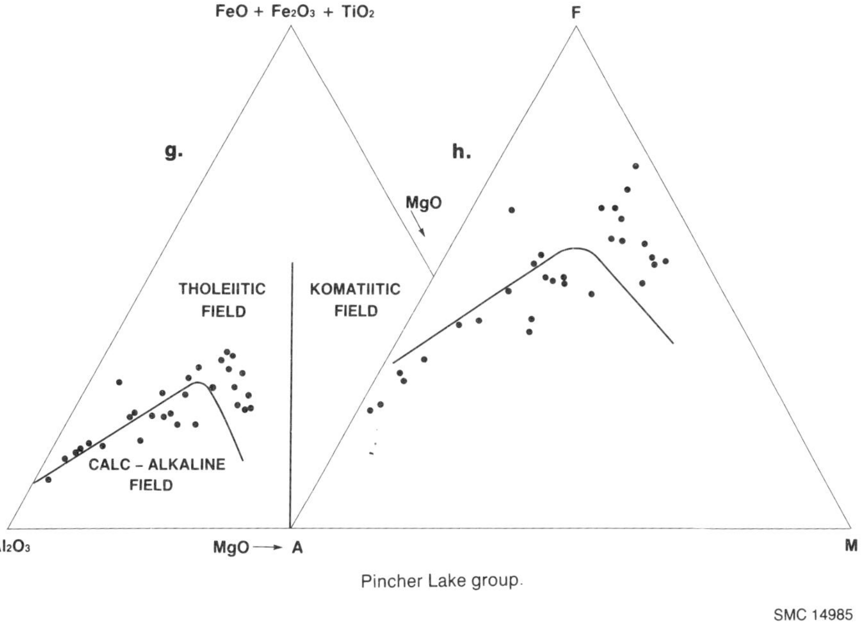

Figure 4g,h. *Pincher Lake group (data from Blackburn 1976, 1979a,b).*

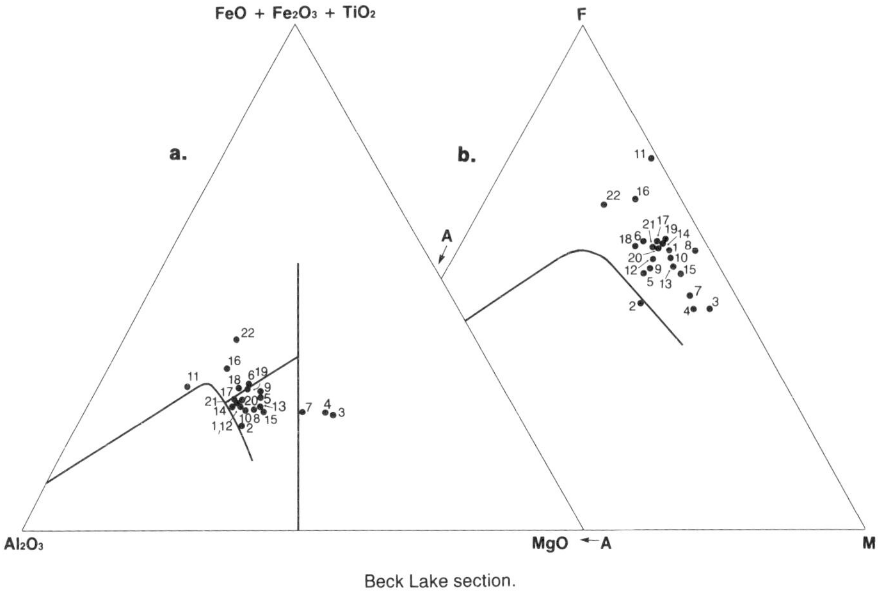

Figure 5—*Chemical classification of metavolcanics. Cation plot after Jensen (1976); AFM plot after Irvine and Baragar (1971). Sample numbers correspond to chemical analyses in Table 4; Beck Lake samples are prefixed with B, Starshine Lake samples with S, Doyle Lake samples with D, and Jonas Lake samples with J.* ***a,b.*** *Beck Lake section of the Wapageisi Lake group.*

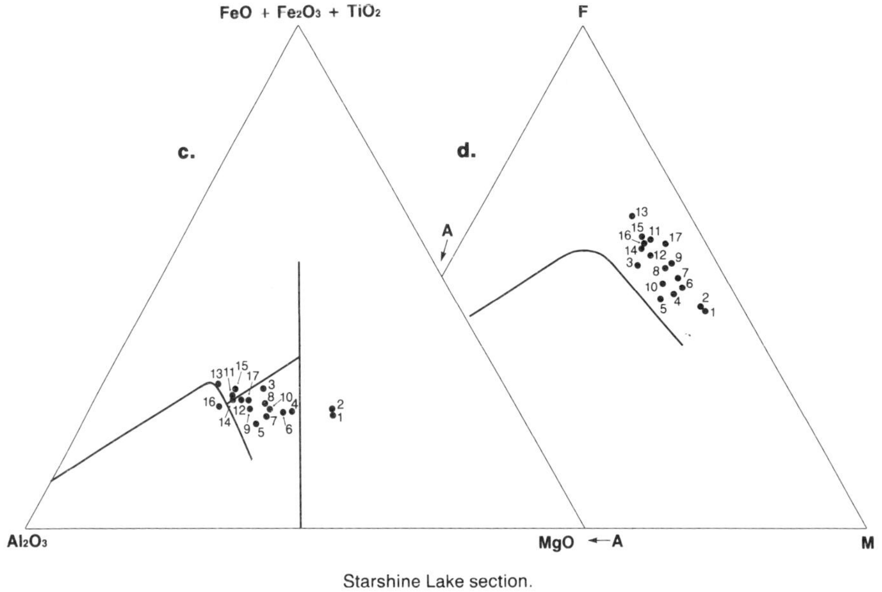

Figure 5c,d. *Starshine Lake section of the Wapageisi Lake group.*

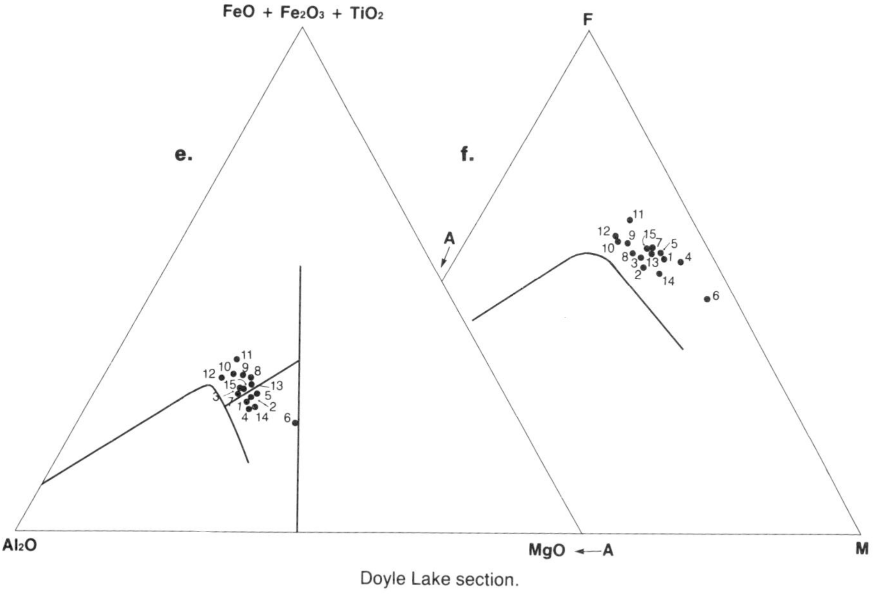

Figure 5e,f. *Doyle Lake section of the Blanchard Lake group.*

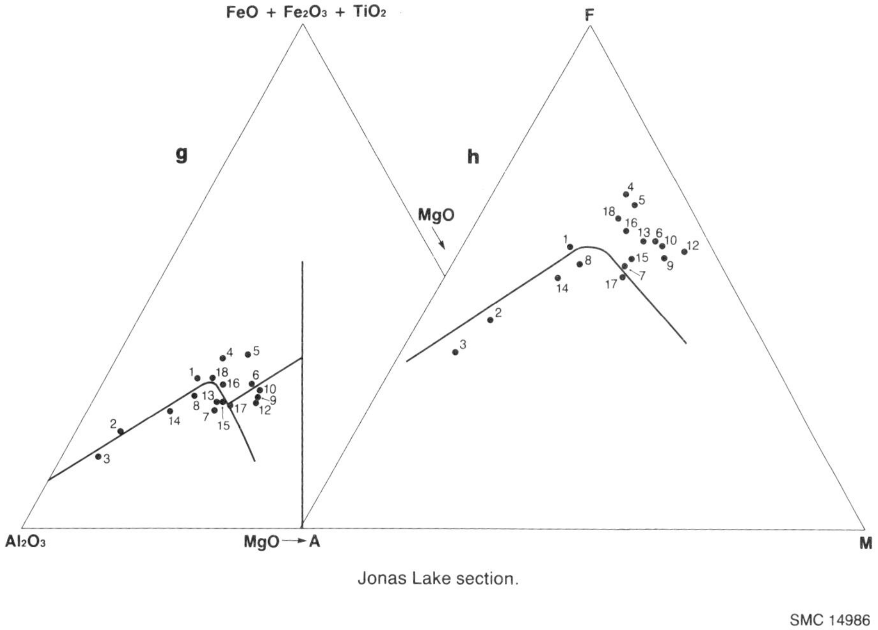

Figure 5g,h. *Jonas Lake section of the Pincher Lake group.*

indicate a komatiitic trend particularly for samples from the lower parts of the group (i.e. samples S1, 2, 4 and 6 in Figure 5c; samples B3, 4, and 7 in Figure 5a; samples 1, 6, 3, and 8 in Figure 4a). The group as a whole plots predominantly on the high magnesium side of Jensen's (1976) empirical divider between magnesium and iron rich suites, and there is a distinct tendency for samples from higher levels of the group to be more iron rich.

The above observations are reflected in variations in composition with stratigraphic height in the Beck Lake Section (Figure 6a) and the Starshine Lake Section (Figure 6b). Both show a distinct decrease in MgO and an increase in Al_2O_3 with height. Variation in Fe_2O_3 is less distinct, but appears to increase with height in both cases. This accounts for the only slight trend toward iron enrichment shown in the cation plots. The distinct increase in TiO_2 with stratigraphic height in both sections has a small but nonetheless positive effect on the iron enrichment trend shown in the cation plot. Alkalies show a rather erratic variation, with no distinct increase or decrease in Na_2O or K_2O in either the Beck Lake or Starshine Lake sections. This is reflected in the AFM diagrams which show no discernable trend toward the A apex. SiO_2 content is in general consistent with basaltic rocks, though samples B2 (Figure 6a) and S3 (Figure 6b) are distinctly and anomalously high, while S8 (Figure 6b) is low. These anomalous SiO_2 values are reflected in anomalous values in most of the other oxides. No clearly defined trend in SiO_2 is evident, though the Beck Lake Section appears to show a slight overall decrease in SiO_2 with stratigraphic height. CaO shows an increase in the Starshine Lake section, but no defined trend in the Beck Lake section. Although the differentiation index (DI) shows no defined trend in either section, the colour in-

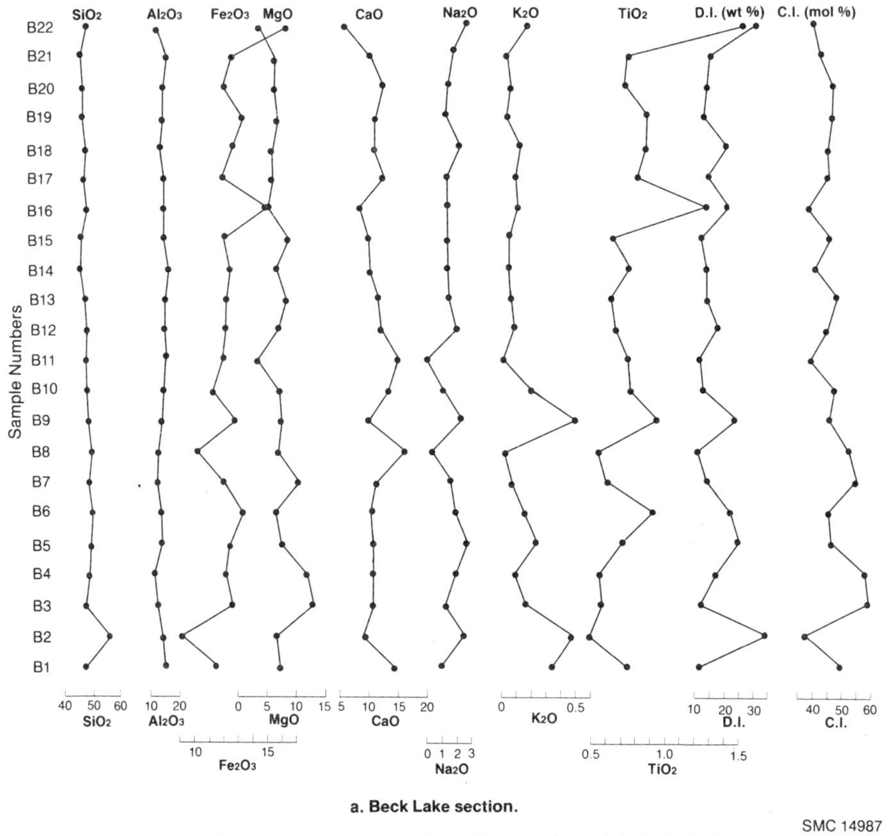

Figure 6—*Variation of chemical composition with stratigraphic height for metavolcanics.*
a. *Beck Lake section of the Wapageisi Lake group.*

dex decreases markedly in both. This is probably a reflection of the fact that the differentiation index, as pointed out by Poldervaart and Parker (1964), does not take into account fractionation in the mafic reaction series.

MANITOU GROUP

During the course of the detailed surveys by the present author the only Manitou group lithologies sampled for petrochemical studies came from the Sunshine Lake trachybasalt formation and associated sills, and the Sunshine Lake felsic porphyry intrusion and an associated flow in the Cane Lake formation. However, Teal (1979) collected samples from all of the formations, 22 of which were analysed by the Geoscience Laboratories of the Ontario Geological Survey in conjunction with the present author's work, and 18 of which were analysed and presented in a thesis by F.J. Longstaffe (1977).

Teal's (1979) and Longstaffe's (1977) data clearly show that both Cane Lake formation and Uphill Lake formation pyroclastics are markedly subalkaline, varying from calc-alkaline andesite to dacite in composition, and samples of both formations cover the same range in compositions. Teal (1979, p.166) pointed out that "even those samples showing evidence of reworking in the alluvial fan environment of the central Uphill Lake Formation ... have a chemical composition very similar to the Cane Lake pyroclastics".

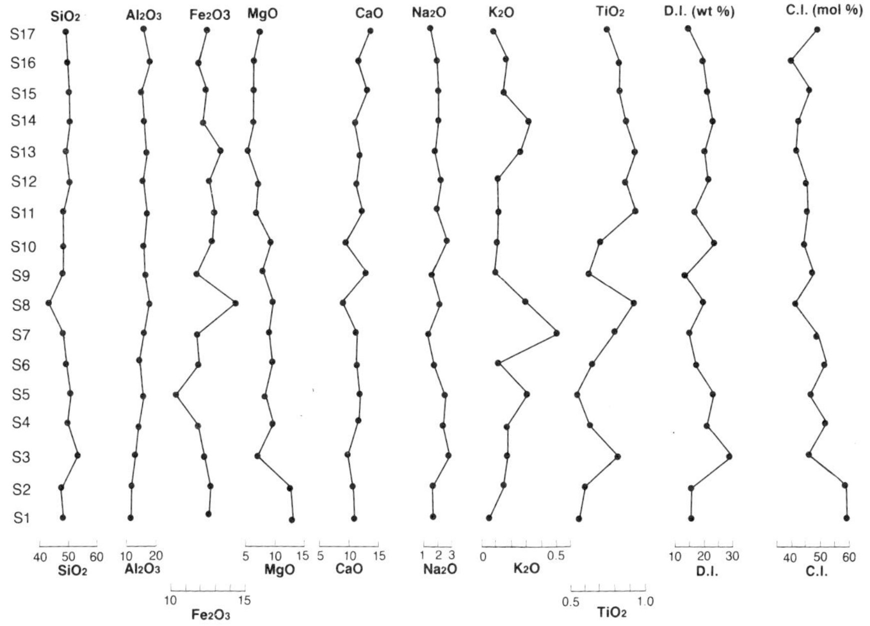

b. Starshine Lake section.

Figure 6b. *Starshine Lake section of the Wapageisi Lake group.*

Teal (1979) discussed the chemistry of argillite and sandstone of the Uphill Lake formation and the Mosher Bay formation. He pointed out that the argillites are of similar composition to Pettijohn's (1957) average shale and slate analyses, "with the exception that the Manitou Group argillites have about twice as much Na_2O as Pettijohn's averages" (Teal 1979, p.169). This he attributed to their derivation from predominantly volcanic material. Sandstone of the Mosher Bay formation, when compared with average Archean greywacke compositions compiled by Pettijohn (1957), are higher in SiO_2 and Na_2O, and lower in Al_2O_3 and MgO (Teal 1979).

Partial chemical analyses of two samples from the Sunshine Lake trachybasalt formation have previously been given by Blackburn (1976, p.36, samples 23, 27) though the formation was originally mapped as a sill. Three other samples given by Blackburn (1976, p.36, samples 24, 25, 26) are in fact intrusive, and interpreted to be a subvolcanic phase of the Sunshine Lake trachybasalt formation. New analyses of these same five samples are given in Table 5. Two other analyses of the same formation were given in Blackburn (1979b, Table 2, samples 16 and 17). In addition, six of Teal's samples analysed by the Geoscience Laboratories of the Ontario Geological Survey are given in Table 5.

These thirteen samples are classified after Irvine and Baragar (1971) in Figure 7. The alkalies vs. silica diagram (Figure 7a) classifies the majority of the samples as alkaline. Two samples are markedly disparate, T45 and 27. In the Ol'-Ne'-Q' diagram (Figure 7b) only three samples fall in the alkaline field. Petrographic descriptions in Blackburn (1976 and 1979b) suggest that the formation as a whole has an alkaline affinity. Potassic feldspar

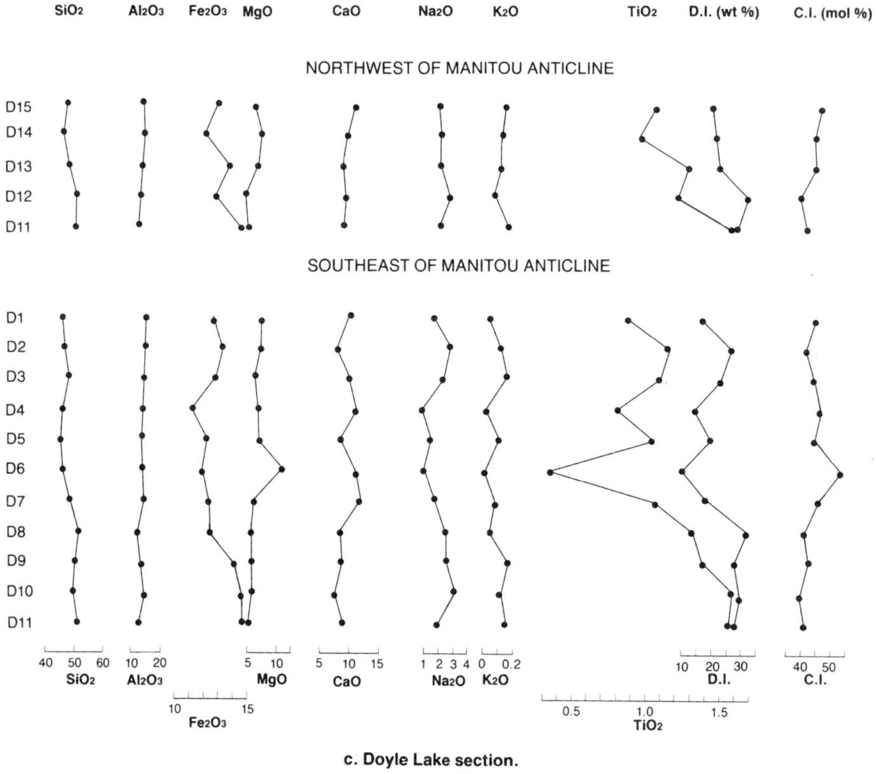

__Figure 6c.__ Doyle Lake section of the Blanchard Lake group.

forms phenocrysts, and matrix textures are similar to trachytic texture. This information arbitrates between the conflicting results of the alkalies-silica plot (Figure 7a) and the Ol'-Ne'-Q' plot (Figure 7b), and suggests that the formation as a whole shows a potassic (Figure 7d), trachybasaltic (Figure 7c) affinity. Because they are widely disparate, two samples, T45 and 27, have been omitted from these latter plots. It is of note that the three intrusive samples, 24, 25, and 26, plot in the same general area as the volcanic rocks, supporting their interpretation as a subvolcanic phase of the volcanics. Furthermore, they are distinctly different (Figures 7a, e) from a metagabbro sample (sample 26 in Blackburn 1979b) of tholeiitic affinity from a body that intrudes the top of the Starshine Lake sub-group south of Sunshine Lake.

Comparison of chemical analyses of a sample from the Sunshine Lake felsic porphyry, and a sample from a felsic flow in the Cane Lake formation (Blackburn 1979b, Figure 5, Table 2, samples 29 and 22 respectively) showed them to be very similar, of calc-alkaline, dacitic composition, thus supporting their common origin. Comparison with fields outlined for samples from the Thundercloud Lake porphyry and associated breccia phase (McMaster 1978) shows them (see Figure 8) to be distinctly different, but to bear closer similarity to the brecciated phase of that porphyry than to the massive phase.

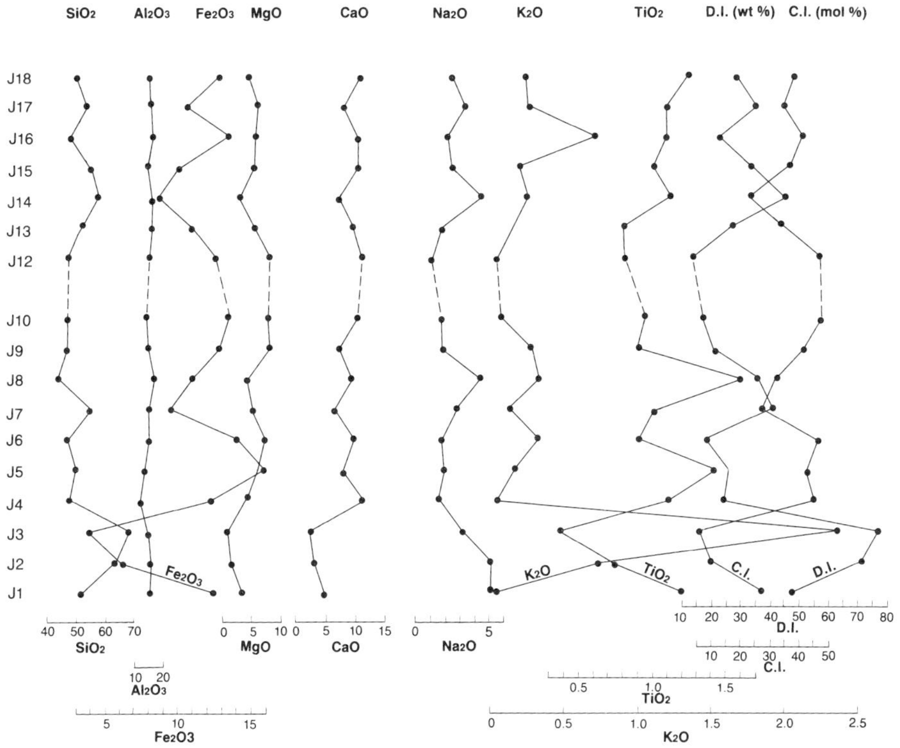

d. Jonas Lake section.

Figure 6d. *Jonas Lake section of the Pincher Lake group.*

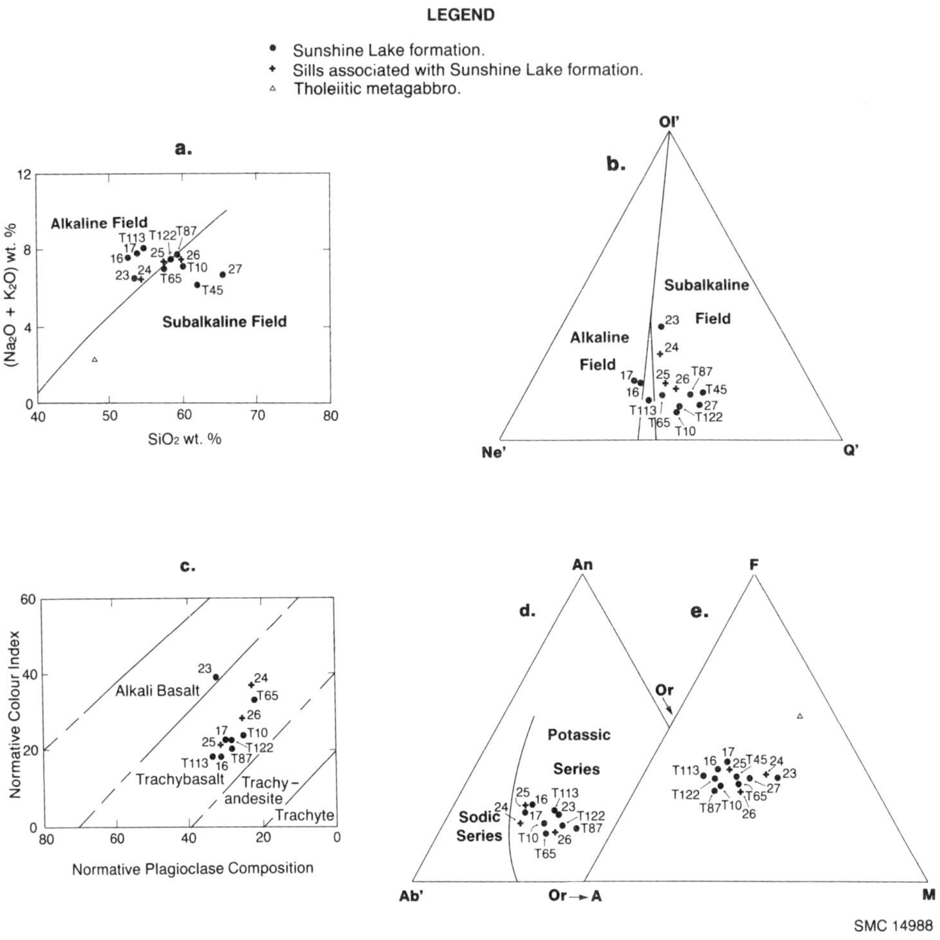

Figure 7—*Chemical classification (after Irvine and Baragar 1971) of samples from the Sunshine Lake formation and associated sills. Samples 16 and 17 are from Blackburn (1979b). Samples 23 to 27 are numbered as in Blackburn (1976). Samples prefixed T are those of Teal (1979).*
a. *Alkalies vs. silica plot.*
b. *Ol'-Ne'-Q' plot. Ol' = normative olivine + 3/4 (enstatite + ferrosilite); Ne' = normative nepheline + 3/5 albite; Q' = normative quartz + 2/5 albite + 1/4 (enstatite + ferrosilite).*
c. *Classification of the potassic suites of the alkaline olivine basalt series. Colour index = normative diopside + enstatite + ferrosilite + magnetite + hematite + ilmenite; normative plagioclase composition = 100 anorthite/(anorthite + albite).*
d. *Normative anorthite-albite-orthoclase plot separating sodic and potassic suite of the alkali olivine basalt series. Ab' = albite + 5/3 nepheline.*
e. *AFM plot. A = weight percent Na_2O + K_2O; F = weight percent FeO + 0.8998 Fe_2O_3; M = weight percent MgO.*

TABLE 5. CHEMICAL ANALYSES (WEIGHT PERCENT) OF SAMPLES[1] FROM THE SUNSHINE LAKE FORMATION AND ASSOCIATED SILLS IN THE MANITOU LAKES AREA. ANALYSES BY GEOSCIENCE LABORATORIES, ONTARIO GEOLOGICAL SURVEY.

Major Elements	**72-B-240[2] (23)**	**72-B-166 (24)**	**72-B-156 (25)**	**72-B-238 (26)**	**72-B-220 (27)**	**T-72-10**	**T-72-45**	**T-72-65**	**T-72-87**	**T-72-113**	**T-72-122**
SiO_2	53.7	54.2	57.5	59.8	65.6	60.0	61.9	57.4	59.2	54.8	58.6
Al_2O_3	13.5	13.5	17.0	14.2	15.8	14.6	15.2	13.4	14.2	17.5	14.6
Fe_2O_3	8.98	8.91	7.57	6.00	5.63	5.70	6.00	6.60	5.50	6.70	6.30
MgO	9.03	7.86	4.33	5.67	3.01	3.90	4.30	5.50	3.90	2.90	3.40
CaO	6.15	5.70	4.57	5.81	2.63	6.00	5.30	7.00	4.30	5.60	5.50
Na_2O	3.08	4.03	4.51	3.66	4.71	3.95	3.42	3.77	3.24	4.04	3.50
K_2O	3.39	2.41	2.75	3.85	2.01	3.45	2.65	3.45	4.42	4.10	4.02
TiO_2	0.90	0.80	0.80	0.80	0.50	0.60	0.80	0.80	0.70	0.70	0.80
MnO	n.d.	n.d.	n.d.	n.d.	n.d.	0.08	0.10	0.10	0.06	0.09	0.09
L.O.I.						1.10	1.32	1.28	4.46	3.29	1.95
Total	98.7	97.4	99.0	99.8	99.9	99.4	101.0	99.3	100.0	99.7	98.8

[1]Sample locations for 72-B series are given in Blackburn (1976, fig. 3). Sample locations for T-72 series are given in Teal (1979, fig. 63).
[2]Sample numbers for 72-B series are given with report sample numbers in parentheses.

STORMY LAKE GROUP

McMaster (1978) has undertaken petrochemical study of samples taken from the Stormy Lake group during the detailed survey by the present author (Blackburn 1979b), and of samples taken during his subsequent additional thesis field work.

Thirteen samples analyzed by McMaster (1978) from the clastic phase of the Stormy Lake group ("Middle Sequence" of McMaster) show the essentially calc-alkaline, andesitic to dacitic, chemistry of these rocks (Figure 8). Because of the limited number and rather

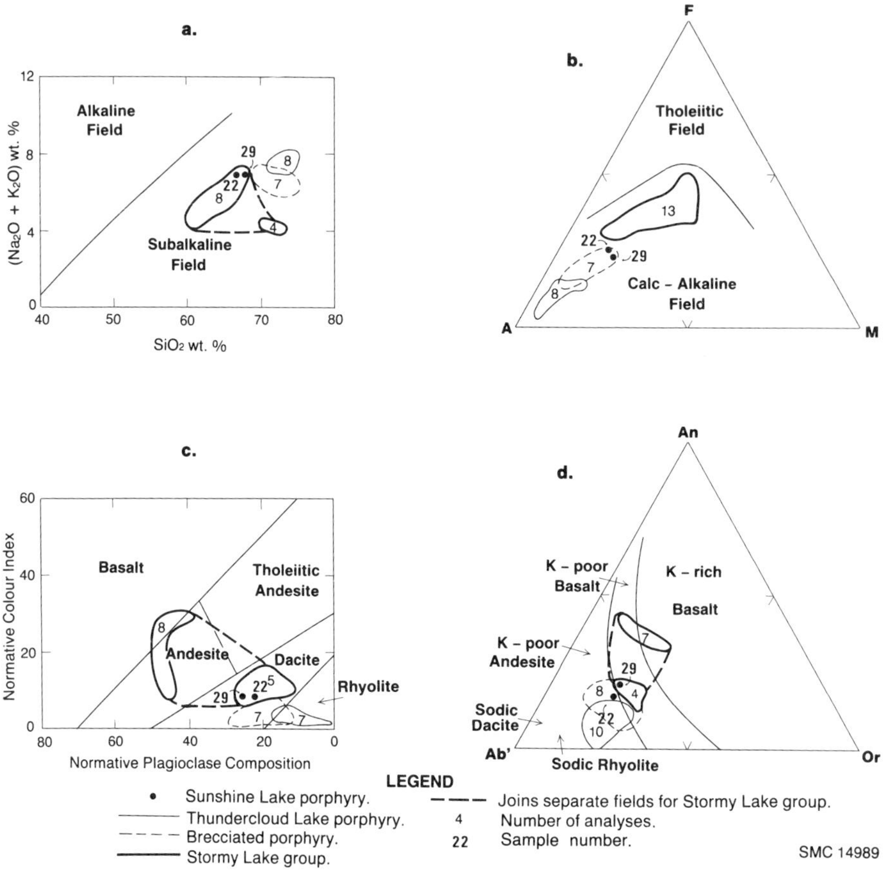

Figure 8—*Chemical variation of samples from subvolcanic felsic porphyries and associated flows, and intermediate pyroclastics of the Stormy Lake group. Fields after Irvine and Baragar 1971 are shown for comparison only. Samples 22 and 29 are from the Sunshine Lake subvolcanic intrusion and a felsic flow in the Cane Lake formation, respectively (from Blackburn 1979b). All other samples are from McMaster (1978).* ***a.*** *Alkalies vs. silica plot.* ***b.*** *AFM plot. A = weight percent $Na_2O + K_2O$; F = weight percent $FeO + 0.8998\ Fe_2O_3$; M = weight percent MgO.* ***c.*** *Colour index = normative diopside + enstatite + ferrosilite + magnetite + hematite + ilmenite; normative plagioclase composition = 100 anorthite/(anorthite + albite).* ***d.*** *Normative anorthite-albite-orthoclase plot. Ab' = albite + 5/3 nepheline.*

heterogeneous nature of the samples, two envelopes are defined for the Stormy Lake group in three of the plots (Figures 8a, c, d).

Eight samples analyzed by McMaster (1978) from the "brecciated porphyry" (felsic tuff and lapilli-tuff of this study) are compared with eleven samples from the massive porphyry. All are calc-alkaline and strongly felsic, though it is noticeable that the "brecciated porphyry" is predominantly dacitic, and the massive porphyry, rhyolitic (Figure 8c). The fields overlap in all four plots, but the area of overlap is small. However, in other binary and ternary variation diagrams presented by McMaster (1978), area of overlap varies from partial to complete, the latter case particularly in plots that involve trace elements. In none of the fifteen such diagrams he presents can any appreciable separation of fields be readily observed. These observations support McMaster's summary statement that "most plots show that the porphyry and brecciated porphyry are chemically identical and can be considered the same rock type" (McMaster 1978, p.97).

BOYER LAKE GROUP

Cation and AFM plots (Figures 4c, d) for samples from the Boyer Lake group show their preponderent tholeiitic chemistry. Two felsic samples are however strongly calc-alkaline. Of the remaining samples, in both the AFM and cation plots most plot in the tholeiitic field and on the high iron side of Jensen's (1976) empirical divider between magnesium and iron rich suites.

The two felsic samples are from a suite of felsic metavolcanics discussed under the heading "Walmsley Creek Felsic Flows and Breccias" in the report on the Boyer Lake - Meggisi Lake area (Blackburn 1979b).

Fifteen samples analysed by McMaster (1975) from a section across the Gabbro Lake sill are shown for comparison on Figures 4c, d. Sample numbers incorporate height in metres from base to top of the sill (e.g. G197 is 197 m from the base of the sill). Their general lower alumina and alkalies ratio compared to Boyer Lake group mafic flows are readily noticeable. Their distinct linear spread on the AFM diagram is interpreted by McMaster (1975) to indicate a tholeiitic, iron enrichment trend. However, there is no consistent increase in Fe/Mg upward in the sill, an observation which is made more evident on the cation plot. These data suggest that the sill, which is taken as representative of other similar gabbroic bodies that intrude the Boyer Lake group, is petrochemically distinct from the mafic flows, and therefore not a subvolcanic equivalent of them.

BLANCHARD LAKE GROUP

Cation and AFM plots (Figures 5e, f) for samples from the Doyle Lake section of the Blanchard Lake group show clearly their tholeiitic chemistry. A majority of samples from the group plot on the high iron side of Jensen's (1976) empirical divider between magnesium and iron rich suites.

The Doyle Lake section crosses the Manitou Anticline at right angles (see Figure 3). The position of the fold axis, interpreted from pillow facings and general arguments of symmetry (Blackburn 1976) is such that sample D11 lies at the core of the fold. Variations in composition with stratigraphic height are shown (Figure 6c) for both limbs of the fold, with sample D11 at the base of both sections. Although only four samples are available for the northwest limb, trends clearly mirror those of the southeast limb. Only the alkalies show no obvious trends. The trends as a whole are not compatible with a normal crystal fractionation series: in fact, most appear to show a reverse trend, viz. decreasing SiO_2; decreasing Fe_2O_3; increasing MgO; increasing CaO; decreasing TiO_2; decreasing DI; increasing CI.

UPPER MANITOU LAKE GROUP

Cation and AFM plots (Figures 4e, f) for samples from the Upper Manitou Lake group show their preponderent calc-alkaline chemistry. Samples from the Frenchman Island Subvolcanic Intrusion show the same range in composition as the volcanic suite, as discussed in the report on the Upper Manitou Lake area (Blackburn 1979a).

PINCHER LAKE GROUP

Cation and AFM plots (Figures 4g, h, 5g, h,) for samples from the Pincher Lake Group show their rather heterogeneous chemistry, ranging from basaltic to rhyolitic, and from tholeiitic to calc-alkaline. On the cation plots, tholeiitic basaltic rocks mostly plot on the high iron side of Jensen's (1976) empirical divider between magnesium and iron rich suites. Also on the cation plots, the more felsic rocks straddle the divider between tholeiitic and calc-alkaline suites.

The heterogeneity referred to above is reflected in the variation in composition with stratigraphic height in the Jonas Lake Section (Figure 6d). No distinct trends are evident, and compositions fluctuate widely.

Granitic Rocks

BATHOLITHIC GRANITIC COMPLEXES

No major petrochemical study of the Atikwa Batholith within the synoptic map-area has been made. Seven samples, of which five are diorite and quartz diorite, and the other two a gabbro and a peridotite, were taken from within the confines of the batholith by the present author (Blackburn 1979a). The diorite and quartz diorite are considered to be hybrid phases. Arguments in favour of this conclusion are based on field observation rather than petrochemistry, and these rocks are not further discussed here.

Meggisi Pluton

Nineteen chemical analyses of samples of the early equigranular granodiorite phase and twenty analyses of samples of the later seriate quartz monzonite phase of the Meggisi Pluton have been presented by Sabag (1979). Although average compositions of these two phases (Table 6) are somewhat similar, iron oxides, MgO and CaO are all appreciably greater, and TiO_2, MnO, and P_2O_5 somewhat greater in the early phase, while SiO_2 and Na_2O are greater in the seriate, later phase. These differences are shown in variation diagrams (oxides vs. SiO_2) by Sabag (1979), which show smooth trends rather than abrupt compositional differences. On a Q-Ab-Or diagram (Figure 9a) the envelope for the later phase is entirely enclosed within the envelope for the early phase, but on an AFM diagram (Figure 9b) the same envelopes are distinct, the envelope for the later phase being closer to the alkali apex.

Also shown on the Q-Ab-Or diagram (Figure 9a) are four samples of porphyry, and on the AFM diagram (Figure 9b), the envelope for these same four samples. On the former, the samples again fall within the envelope for the early granitic phase, while on the latter the envelope is again distinct, and closer again to the alkali apex.

TABLE 6. AVERAGE CHEMICAL COMPOSITIONS OF GRANITIC AND PORPHYRY BODIES IN THE MANITOU LAKES AREA.

Body	Meggisi Pluton and porphyries			Taylor Lake Stock		Scattergood Lake Stock	Carleton Lake Stock	Thundercloud porphyry and breccia	Sunshine Lake porphyry
Rock Types	Equigranular granodiorite (early phase)	Seriate quartz monzonite (late phase)	Porphyry	Monzonite and syenodiorite	Quartz monzonite and granodiorite	Quartz monzonite	Quartz monzonite	Porphyry and breccia	Porphyry
No. of Samples	19[2]	20[2]	4[2]	7[1]	11[1]	25[1]	1[2]	19[2]	1[2]
SiO_2	68.37	72.01	72.55	59.35	68.99	68.39	67.50	73.08	67.90
Al_2O_3	15.25	15.27	15.18	15.77	15.85	15.64	16.00	15.14	15.10
Fe_2O_3	1.18	0.67	0.55	6.08	2.31	2.75	1.54	1.36	1.41
FeO	1.50	0.59	0.39	—	—	—	1.03	0.23	1.46
MgO	1.63	0.39	0.30	3.00	1.29	1.50	0.78	0.93	1.97
CaO	2.99	1.80	1.53	5.50	2.38	2.85	2.37	1.98	2.97
Na_2O	4.43	5.09	5.35	5.55	5.10	5.08	5.02	4.84	4.58
K_2O	2.62	2.75	2.41	3.66	3.58	3.27	3.66	2.20	2.35
TiO_2	0.36	0.18	0.12	0.54	0.29	0.29	0.28	0.21	0.37
P_2O_5	0.09	0.03	0.02	0.40	0.14	0.17	0.14	0.07	0.13
MnO	0.03	0.02	0.02	0.14	0.05	0.05	0.06	0.04	0.05
Source	Sabag (1979)			Pichette (1976)		Birk (1978)	Blackburn (1976)	McMaster (1978)	Blackburn (1979b)

[1]Averages of analyses normalized to 100% anhydrous.
[2]Averages of raw analyses.

INTRABELT PLUTONS

Taylor Lake Stock

Eighteen chemical analyses of samples of major phases from the Taylor Lake Stock have been presented by Pichette (1976). When grouped into two major groupings (monzonite and syenodiorite; quartz monzonite and granodiorite), average compositions (Table 6) are appreciably different: Fe_2O_3, MgO, and CaO are considerably greater, and TiO_2, MnO, Na_2O, and P_2O_5 are somewhat greater in the monzonite and syenodiorite, while SiO_2 is considerably greater in the quartz monzonite and granodiorite. These differences are shown in variation diagrams (oxide vs. differentiation index) by Pichette (1976), which show smooth trends rather than abrupt compositional differences.

On a Q-Ab-Or diagram (Figure 9a) there is distinct separation of the two envelopes for these major groupings, the envelope for seven monzonites and syenodiorites being along the albite-orthoclase edge, and that for 11 quartz monzonites and granodiorites more cen-

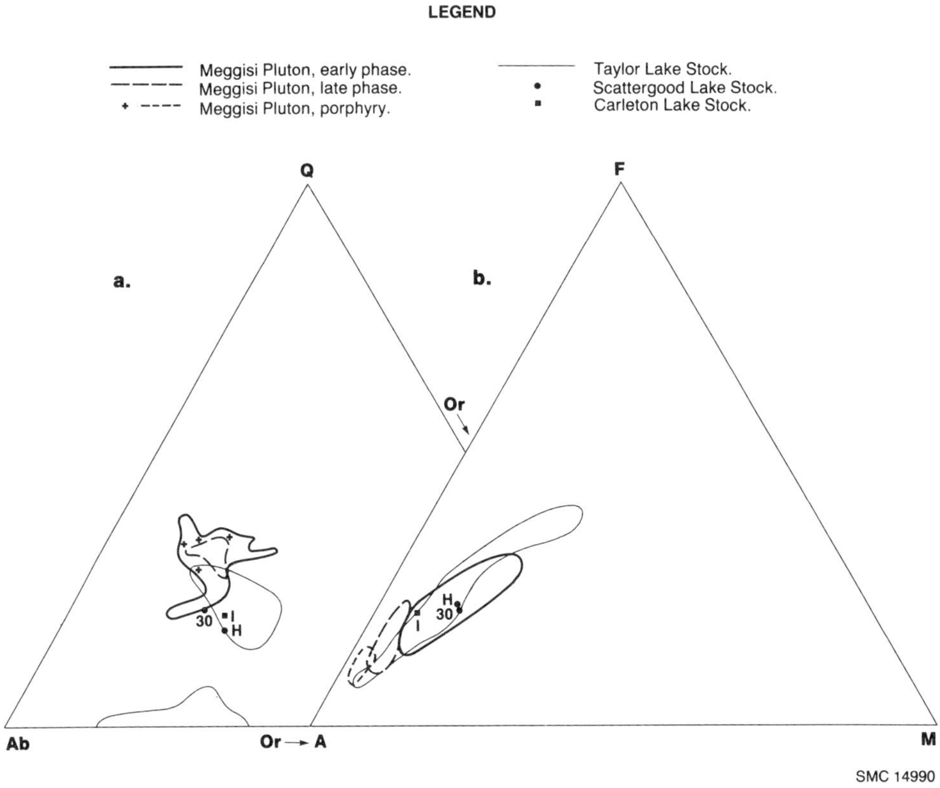

Figure 9**—Chemical variation in granitic rocks and associated felsic porphyries. Meggisi Pluton samples (19 samples from early phase, 20 samples from late phase) from Sabag (1979). Taylor Lake Stock samples (18) from Pichette (1976). Scattergood Lake Stock samples (2) and Carleton Lake Stock sample (1) from Blackburn (1976, 1979b).* ***a. *Normative quartz-albite-orthoclase plot.* ***b.*** *AFM plot. A = weight percent $Na_2O + K_2O$; F = weight percent $FeO + 0.8998\ Fe_2O_3$; M = weight percent MgO.*

trally located. On an AFM diagram (Figure 9b) the 18 samples define a somewhat linear envelope, varying from quartz monzonite and syenodiorite near the alkali apex to monzonite and syenodiorite toward the iron-magnesium side.

Scattergood Lake and Carleton Lake Stocks

Twenty-five chemical analyses of samples from the Scattergood Lake Stock have been presented by Birk (1978), and two by Blackburn (1976, 1979b). The averages given by Birk (1978) of his 25 samples are given here in Table 6. Birk (1978) pointed out the remarkable chemical uniformity of the values for the 25 samples. Also remarkable is the chemical similarity of the Carleton Lake Stock (Table 6). On Q-Ab-Or and AFM diagrams (Figure 9), the two samples from Scattergood Lake and the one from Carleton Lake collected by the present author (Blackburn 1976, 1979b) plot close together.

METAMORPHISM

Field observations and subsequent thin-section studies carried out during the detailed mapping (Blackburn 1976, 1979a,b) suggest that grade of metamorphism within the supracrustal and subvolcanic rocks varies from greenschist to amphibolite facies. In general, greenschist facies characterises the more internal parts of the metavolcanic-metasedimentary belt, while amphibolite facies characterises the margins. Because lithologies near the margins of the belt are predominantly basaltic and therefore insensitive to minor changes in metamorphic conditions, and because in general lithologic boundaries appear to parallel the edges of the belt, subtle changes in metamorphic grade were not noted. Field observations led to the establishment of a tentative boundary between "greenschist" and "amphibolite" on the northwest side of the belt. This boundary (see Figure 10, Chart B, back pocket) crosses the northwest limb of the Manitou Anticline, and at one point may also cross the axial plane trace. Grade of metamorphism is observed to be related to proximity to the enclosing batholiths, and therefore is interpreted to be imposed by emplacement of the batholiths. The pervasive greenschist facies metamorphism in the internal parts of the belt may be due to batholith emplacement, or a regional event or events of unknown origin.

The post tectonic Taylor Lake, Scattergood Lake, and Carleton Lake Stocks appear to have had only minor effect, imparting a contact metamorphism, as evident in upgrading of greenschist to amphibolite at some places near their contacts.

STRUCTURAL GEOLOGY

In the preceding sections of this report dealing with stratigraphy and petrochemistry of the Manitou Lakes synoptic area, structure of the metavolcanic-metasedimentary belt has been implied. In this section structural features are discussed in so far as they relate to an interpretation of the stratigraphy, and will be applied in the following section to an interpretation of the geological evolution of the map-area.

Major Faults

A number of major faults have been mapped in the area, but two of these are of critical importance in making stratigraphic correlation, and in interpreting the general geologic evolution of the area.

MANITOU STRAITS FAULT

Thomson (1933, p.4) first recognized that "a fault of considerable magnitude extends through the Manitou straits and northward to Kabagukski Lake", and possibly southward beneath Lower Manitou Lake. The present author has confirmed its presence (Blackburn 1976, 1979a,b), and traced it beyond Kabagukski Lake. Subsidiary, parallel, faults have also been recognized, particularly in the southwest (Blackburn 1976).

Amount and direction of movement on the fault remains equivocal, and is probably different at different localities along the fault based on observed variation in offset of lithologic units. At the northeast end, lithologies of the Boyer Lake group, and the axial plane of the Kamanatogama Syncline, are truncated at a high angle by the fault. Southwest of Mosher Bay the fault lies parallel to strike, while at Mosher Bay itself, minor folding in the Mosher Bay formation has variously been interpreted as related to movement along the Manitou Straits Fault (Thomson 1933), or as related to movement along the Mosher Bay-Washeibemaga Lake Fault (Blackburn 1979b).

Correlation of stratigraphic units across the fault is hampered by lack of suitable marker beds. General successions on either side of the fault suggest correlations across it, discussed below under "Geologic Evolution of the Manitou Lakes Area". Features close to and within the fault zone that pertain to correlation are discussed here.

A distinctive unit of magnetite ironstone interbedded with sandstone and siltstone, occurs within the Mosher Bay formation. At Beaverhead Island, this unit appears to lie within the fault zone (Blackburn 1976, p. 35). Aeromagnetic data, from Map 1153G (ODM-GSC 1961), summarized in Figure 9 of Blackburn (1976), and from a more detailed airborne survey carried out for Daering Explorers Corporation Limited in 1967 (Assessment Files Research Office, Ontario Geological Survey, Toronto; Blackburn 1976, p.63), suggest that the unit continues beneath the waters of Lower Manitou Lake, and possibly on the northwest side of the fault, as shown on Figure 9 of Blackburn (1976). If this latter conclusion is true, then at this locality movement along the fault is minimal. Facing of tuffs and pillowed lavas on a small island in the bay on the south side of Beaverhead Island is to the southeast (Map 2476, back pocket), whereas the Mosher Bay formation metasediments on the southeast side of the fault face northwest. A tight, sheared-through syncline is therefore a possibility at this locality. However, although metasediments are present near Beaverhead Island to the northwest of the fault, no ironstone, either interbedded with metasediments, or as pebbles in conglomerates, has been found anywhere in outcrop northwest of the Manitou Straits Fault.

MOSHER BAY-WASHEIBEMAGA LAKE FAULT

The Mosher Bay-Washeibemaga Lake Fault has been discussed previously (Blackburn 1979b). In summary, evidence for the existence of the fault consists of 1) angular discordance between the Boyer Lake group and Manitou and Stormy Lake groups; 2) intense shearing in rocks along the contact between these groups; and 3) tight mesoscopic folds within the Manitou and Stormy Lake groups, but only close to their contact with the Boyer Lake group.

Movement along the fault is interpreted to have occurred concommitantly with that along the Manitou Straits Fault, thus accounting for the variable amount of displacement along the latter. The Mosher Bay-Washeibemaga Lake Fault is interpreted to be a thrust fault, along which the Boyer Lake group was thrust over the Manitou and Stormy Lake groups. If this interpretation is correct, then the portion of the Manitou Straits Fault north of Mosher Bay may have acted as a dextral tear fault, along which the Boyer Lake group was displaced horizontally southward relative to the Pincher Lake group.

Major Folds

MANITOU ANTICLINE

The existence of an anticlinal fold structure northwest of the Manitou Straits Fault was first recognized by Goodwin (1965), who named it the Manitou Anticline (Goodwin 1970, Table 2). Subsequent mapping by the present author (Blackburn 1976, 1979a, b) has confirmed its presence, and more clearly indicated its extent and form.

Primary and tectonic structural data collected during the detailed surveys are condensed and combined in Figures 10, 11, and 12 (Chart B, back pocket). All measured bedding attitudes are indicated in Figure 10. Bedding is comparatively abundant at Upper Manitou Lake, and near Beaverhead Island, predominantly within the Upper Manitou Lake group. On the northwest limb of the fold, in the vicinity of Merrill Lake, amphibolite grade of metamorphism has imposed a gneissosity. However, as discussed previously (Blackburn 1976), the gneissosity in general can be shown to parallel bedding and stratigraphic contacts where the two are observed together, and it can be assumed that in bedded pyroclastics in this part of the map-area at least, attitude of gneissosity generally reflects that of bedding and stratigraphic contacts. On this basis two cross-sections (AB, CD) have been constructed. They clearly show the fold to be overturned, and that the degree of overturning is greater in the southwest. Closure of the fold is suggested above present erosional level along section CD, but not along section AB. Section CD is close to a portion of the fold where closure can be seen in the horizontal plane, and where it is believed the fold plunges to the northeast or east, on the basis of minor and mesoscopic folds (Blackburn 1979a, p.54); section AB on the other hand crosses the fold where no closure can be demonstrated in the horizontal plane, and where stratigraphic markers (Map 2476, back pocket) parallel and are not folded around the fold axial plane trace. It may be inferred that in the vicinity of section AB, because of the lack of fold closure, the fold axis cannot plunge directly down the dip of the primary layering. It may further be inferred, because of the form of the fold (i.e. parallel limbs) and the evidence that further to the northeast it plunges northeastward, that at this point it plunges at a shallow angle to the northeast.

Further to the northeast, where the fold axial plane trace passes through the Pincher Lake group, bedding, where obtainable for measuring, is steep and predominantly vertical. Insufficient information is available to construct a cross-section of the fold in this area.

All foliations shown on Map 2476 (back pocket) are shown on Figure 11: dip values for these foliation data points have been contoured. The foliations chosen are representative of those measured during the detailed survey. Because of lack of information in some parts of the area, spurious contours may have been interpreted, particularly where "bull's-eyes" result. If these "bull's-eyes" are disregarded it will be seen that foliation dip values are steep over the northeast part of the fold, and decrease systematically toward the southwest, with dips consistently to the southeast.

All lineations measured during the detailed survey are shown on Figure 12. Within granitic terrain, lineations recorded were mineral lineations, defined by elongate amphiboles and feldspars, quartz rodding, and mica trains. Within supracrustal rocks mineral lineation was found to lie parallel or nearly parallel to stretching of clasts, so that in practice both mineral lineation and the attitude of clast long-axes were measured. On Figure 12, plunge values for all these lineation data points have been contoured. Disregarding local observations, plunge values are steep to vertical in the vicinity of the Manitou Straits Fault, and are also steep over the northeast part of the fold, while they decrease systematically toward the southwest. A distinct fanning of the lineations is evident, and in all cases they plunge away from the granitic rocks in the northwest.

When Figures 10, 11, and 12 are compared, certain relationships emerge. In the field, lineations were in general observed to lie in the plane of the foliations, where the two were observed together, and this relationship is supported by Figures 11 and 12. Within the supracrustal domain lineations plunge steeply in the plane of the steep foliation close to the

Manitou Straits Fault, and also in the north part of the domain. Toward the west, lineations plunge moderately to shallowly down the dip of the foliation where it also is moderately to shallowly dipping, and within the strike of the foliation where it is steeply dipping.

The relationship between primary features such as bedding and stratigraphic contacts, and tectonic structures is less clear. Foliation mostly is parallel or sub-parallel to bedding, except in the nose of the Manitou Anticline. It therefore appears to be axial plane foliation, developed during the major folding event. If the interpretations regarding plunge of the axis of the Manitou Anticline are correct, then lineation does not parallel this axis throughout the fold structure, and particularly in the southwest. It would appear that lineation is a superposed, extensional, feature, probably less related to the major folding of the Manitou Anticline than to the emplacement of the Atikwa Batholith. This interpretation fits with the concept that metamorphism is related to batholith emplacement, because lineation is defined both by elongation of clasts and minerals, and by growth of metamorphic minerals.

KAMANATOGAMA SYNCLINE

A major synclinal fold southeast of the Manitou Straits fault and north of the Mosher Bay – Washeibemaga Lake thrust fault was first interpreted by Goodwin (1965), following the view of Thomson (1933) that the Manitou-Stormy Lake belt was essentially synclinorial. Goodwin later (1970, Table 2) applied the name Mosher Bay Syncline to it. Mapping (Blackburn 1979b) and regional reconnaissance (Blackburn *in* Trowell *et al.* 1977) by the author has confirmed the presence of a syncline, but because its position is not as suggested by Goodwin (1965, 1970), and does not pass near Mosher Bay, Goodwin's name is not retained. The name Kamanatogama Syncline is applied here after a small lake at the east margin of the present map-area; the synclinal axis lies immediately south of it.

Reconnaissance by the author (Blackburn *in* Trowell *et al.* 1977) has established that the axial plane trace of the syncline extends at least as far east as Noxheiatik Lake, a distance of 5 km east of the map-area, and regional structural considerations (Blackburn *in* Trowell *et al.* 1977) suggest that it may continue southeastward toward Bending Lake at the southeast end of the Manitou-Stormy Lakes belt (Blackburn *et al.* 1981).

Because of lack of bedding in the folded strata, only indirect information is available on the plunge of this fold and the dip of the axial plane. The inferred length of the axial plane trace suggests that the fold plunges shallowly, and to the west, because of the increase in width of strata in this direction. Also, in the vicinity of Walmsley Lake and the south part of Boyer Lake, pillow tops (Blackburn 1979b) are inconsistent, suggesting a shallow dip of the flows in this area. Conversely, near Kamanatogama Lake (Blackburn 1979b), steep attitudes of stretched pillows suggest a tight fold structure, that is compatible with either a shallowly or steeply plunging fold axis. The steep pillow attitudes also suggest that the axial plane at this locality is steep, and therefore may be steep throughout the fold.

Offset of the Kamanatogama Syncline by the Taylor Lake Fault and another fault between it and the Manitou Straits Fault is critical in an understanding of the geological evolution of the area. This faulting predated emplacement of both the Gabbro Lake and Mountdew Lake gabbro intrusions, thus establishing a major period of folding prior to emplacement of these gabbros. Because McMaster (1975) has shown the Gabbro Lake body to be a sill that must have been emplaced horizontally (see "Gabbro Lake, Mountdew Lake, Mosher Bay, and Kenny Lake Gabbros", this report), it follows that at least at its western end, the axial plane of the Kamanatogama Syncline was not vertical at time of folding, and may have been near to horizontal. Such an interpretation is compatible with the interpretation of the Mosher Bay – Washeibemaga Lake Fault being a low-angled thrust.

C.E. BLACKBURN

GEOLOGIC EVOLUTION OF THE MANITOU LAKES AREA

The following account summarizes current evidence and data from within the synoptic area, including geochronologic data (Davis *et al.* 1980). To date, available geochronology supports and augments the interpretation given here: further geochronological work in progress is designed to further test some of the hypotheses.

Earliest events were eruption of predominantly tholeiitic basalts of the Wapageisi Lake and Blanchard Lake groups on an undetermined basement. Boyer Lake group basalts, also tholeiitic, but more iron rich were probably erupted in part contemporaneously, an unknown distance to the present north of the former two groups. The Boyer Lake group was probably erupted over a much longer interval, as indicated by a 2704 m.y. date (Davis *et al.* 1980) obtained from felsic rocks near the top of the sequence and which is younger by 85 m.y. than a date obtained from the Thundercloud Lake porphyry which intrudes the top of the Wapageisi Lake group. In excess of 7000 m of predominantly basaltic flows and subvolcanic gabbros were emplaced, probably over a wide area.

Emplacement of the early phase of the Meggisi Pluton into the lower units of the Wapageisi Lake group probably commenced well before eruption of the thick, basaltic sequence had terminated, and the late, seriate phase was emplaced toward the end of eruption of the basalt platform. Amphibolite grade metamorphism was imposed at this time by the pluton in basaltic rocks adjacent to its margin.

There is abrupt change to calc-alkaline volcanism at the top of the Blanchard Lake and Wapageisi Lake groups. At Upper Manitou Lake, the Frenchman Island Subvolcanic Intrusion is probably the vent-filling for much of the pyroclastic volcanism in that group, while vents were probably located where the Sunshine Lake Subvolcanic Intrusion is emplaced at the base of the Manitou group, and where the Thundercloud Lake Subvolcanic Intrusion (dated at 2789 m.y. Davis *et al.* 1980) is emplaced at the base of the Stormy Lake group. A porphyry body emplaced within the Cane Lake formation, southwest of Cane Lake, may also represent a vent-filling. Teal (1979) has presented evidence based on grain orientation analysis that a further vent lay in the vicinity of Cane Lake, but above present erosion level. Rhyolitic flows at Sunshine Lake and west of Washeibemaga Lake are probably eruptive phases of the subvolcanic porphyry bodies in those areas. In turn, the subvolcanic porphyries are probably high-level phases of the late, seriate phase of the Meggisi Pluton, in accord with a model for this type of igneous association presented by Elbers (1976).

Moderately alkalic, trachybasaltic flows and associated sills were erupted toward the close of calc-alkalic pyroclastic volcanism in the Manitou group and possibly the Stormy Lake group, but did not extend into the Upper Manitou Lake group. These flows were erupted subaerially, as were much of the calc-alkalic pyroclastics.

The sedimentary sequence that developed above the Sunshine Lake trachybasalt formation initially was supplied with detritus from emergent volcanic, mainly dacitic, piles, and later from unroofed granitic sources, possibly including the Meggisi Pluton. Teal (1979) has shown that the sedimentary sequence is characterised by the development of an initial alluvial fan and braided fluvial system, followed by a later submarine fan environment.

The development, at the top of the Manitou group, of a submarine fan environment indicates tectonic instability and subsidence, possibly along an axis parallel to, but not necessarily coincident with, the location of the present-day Mosher Bay – Washeibemaga Lake Fault.

The Pincher Lake group, a mixed tholeiitic to calc-alkaline, flow to pyroclastic sequence was erupted above the Upper Manitou Lake group. No equivalent group is present above the Manitou and Stormy Lake groups, probably because volcanism had ceased in this area, where submarine sedimentation was proceeding.

Commencement of emplacement of granitoid batholithic rocks in the vicinity of the

Atikwa Batholith probably accounts for tight folding about the Manitou Anticline. Concommitantly, horizontal shortening in a north-south direction led to initial tilting of the Manitou group, folding about the Kamanatogama Syncline within the Boyer Lake group, probably still located some distance to the present north, and then overthrusting of the Boyer Lake group along the Mosher Bay – Washeibemaga Lake Fault. The Manitou Straits fault must also have been active at this time, and possibly prior to this, at intitiation of folding of the Manitou Anticline. The Gabbro Lake, Mountdew Lake, Mosher Bay, and Kenny Lake Gabbros were emplaced into the Boyer Lake group probably during this horizontal shortening episode.

Further development of the Atikwa Batholith tightened the Manitou Anticline, and also imposed amphibolite grade metamorphism on supracrustal rocks adjacent to its margin.

Peak grade of metamorphism adjacent to the batholith was reached at this time because the amphibolite-greenschist isograd cross-cuts the axial plane trace of the Manitou Anticline.

Continuing north-south horizontal shortening tightened folds, particularly the Kamanatogama Syncline, and steepened the Manitou group sequence, and possibly high-angled thrusting continued along the Mosher Bay – Washeibemaga Lake Fault.

Present geochronologic data indicate that horizontal shortening terminated before, at the latest, 2678 m.y. (Davis *et al.* 1980), a date obtained from the Taylor Lake Stock. This stock, and probably also the Scattergood Lake Stock, was emplaced following movement on the Mosher Bay – Washeibemaga Lake Fault, which it cross-cuts.

The latest faulting took place along north-northeast trending faults: most of the movement is sinistral, as seen in the Taylor Lake Fault. At its northern end, movement on the fault has dextrally offset the Kamanatogama Syncline, and because the Gabbro Lake sill is unaffected by this faulting it must have been an early movement, along which the Taylor Lake Fault was reactivated.

The latest intrusion was that of the predominantly northwest-trending diabase dikes, probably later than the major movement on the north-northeast trending faults, because one of them crosses the Taylor Lake Fault without offset, near Scattergood Lake. Near Cane Lake, another dike has however been dextrally offset along the same fault set.

Cratonisation before emplacement of the diabase dikes suggests that subsequent geologic history has been passive since Precambrian time, apart from deep erosion and peneplanation.

ECONOMIC GEOLOGY

Accounts of mineral deposits, their relationships to geological features, descriptions of properties and exploration, and recommendations for further mineral exploration, have been given in preceeding reports on detailed mapping by the author (Blackburn 1976, 1979a, b). These include discussions of the following commodities: gold, copper, zinc, nickel, molybdenum and iron.

Discussion is limited here to an assessment of genesis of the four major types of deposits found, and the role this can take in exploration where warranted.

Gold Deposits

Gold has historically been the cause of interest in the Manitou Lakes and surrounding region. Current prices for gold make the area once again important as an area of exploration for this commodity.

Following a period of reconnaissance mapping through the metavolcanic-metasedimentary belt extending from Manitou Lakes to Lake of the Woods, Thomson (1936) presented a summary of the gold deposits, and emphasized that (p.699):

"The two features essential to all the deposits seem to be: (1) granite or porphyry intrusives to act as the source of the mineralizing solutions... and (2) a means of access to these solutions." He was reiterating the then current theories of hydrothermal emplacement in structurally favourable traps.

A study of geographic distribution of known gold occurrences and their relationship to the stratigraphy of broad geologic units and to shear zones in the Lake of the Woods – Manitou Lakes – Wabigoon region led Goodwin (1965) to the conclusion that gold is preferentially located in two stratigraphic zones: a sequence of felsic metavolcanics, and subjacent mafic metavolcanics. Although not stated, the implication can be made that a volcanogenic origin, unrelated to structural controls along shear zones, was favoured.

In a discussion of mineral exploration targets in northwestern Ontario, Riley *et al.* (1971) pointed out that 30 percent of gold deposits, in a far broader region than that considered by either Thomson (1936) or Goodwin (1965), occur in epizonal felsic intrusions that form only 8 percent of the metavolcanic-metasedimentary belts. They did not substantiate Goodwin's (1965) finding that felsic metavolcanics and subjacent mafic metavolcanics are preferentially favourable, but did conclude that the epizonal felsic intrusions are (p.38) "probably a concomitant intrusive phase of felsic volcanism and represent the subvolcanic equivalent of the felsic metavolcanics. The gold deposits are thus related to the volcanic process but appear to have been concentrated in the magmatic rather than extrusive phase of felsic volcanism".

Mackasey *et al.* (1974), following a literature study of 42 past-producing mines in the Wabigoon Belt or Subprovince, observed that although 18 of the mines were located in mafic metavolcanics, total gold production from them was negligible compared with that from 10 mines that were located in metasediments, in many cases that contained iron formation. This led them to the conclusion that sources for gold include, among others, (1) mafic volcanic and associated mafic intrusive rocks, and (2) the sedimentary environment in which some iron formations form; the felsic intrusions acted as (p.10) "sources of heat, volatiles, water, and silica that extracted gold from the surrounding volcanic (and sedimentary) rocks".

Beard and Garratt (1976), in compiling data on gold deposits of the Kenora-Fort Frances area, noted the close association of many deposits with felsic intrusive rocks, and with mafic metavolcanics. The former they ascribe to a genesis akin to that proposed by Riley *et al.* (1971), while suggesting, as did Mackasey *et al.* (1974), that mafic volcanic rocks are also a major source.

Trowell *et al.* (1980), in a preliminary account of a regional study of part of the Wabigoon Belt, have reiterated the association of many deposits with a mafic volcanic source, suggesting that, as in the Timmins area (Pyke 1975), it is the more ultramafic rocks of this type that will host or be the source of gold for major deposits. However, they also noted the association of many deposits with subvolcanic, or epizonal, felsic intrusions, and granitic batholithic phases that may be subvolcanic sources of felsic volcanism, thus supporting the conclusions of Riley *et al.* (1971) noted above.

Table 7 groups gold deposits in the Manitou Lakes area, and summarizes their association with geologic features, and interpreted genesis by previous authors and the current author. From the table it is obvious that no single genetic model can account for all of the gold deposits. In the case of those near Gold Rock, at least three alternative models have been presented and in consideration of the various observations by workers since Thomson's initial review, other models might be presented for some of the other deposits in the map-area. For example, deposits northwest of the Manitou Straits Fault appear to agree with the model presented by Goodwin (1965), in that they could be considered to lie within two stratigraphic zones: a sequence of felsic metavolcanics (Upper Manitou Lake group and lower part of the Pincher Lake group) hosting the deposits at Gold Rock and Upper Manitou Lake; and subjacent mafic metavolcanics (Blanchard Lake group) hosting deposits between Manitou Island and Rector Lake.

The latter association of deposits with the Blanchard Lake group appears to support

TABLE 7. ASSOCIATION AND GENESIS OF GOLD DEPOSITS IN THE MANITOU LAKES AREA.

Location	Deposit	Association of Gold	Genetic Interpretations
Gold Rock	Big Master mine Laurentian mine Elora (or Jubilee) mine Selby Lake prospect Paymaster prospect Little Master prospect Volcanic Reef (or Vulcan) prospect Victory (or Upper Neepawa) occurrence Detola prospect	In quartz veins spatially associated with felsite units.	Thomson (1942): structural control; two types present (a) quartz masses in the Jubilee "break" (hydrothermal), (b) quartz masses in fractures in felsite dikes. Blackburn (1979b): Volcanic; felsites are flows or sills, acidic differentiates in cyclic volcanism. Trowell *et al.*(1980): sea-floor alteration/ metamorphism or hydrothermal alteration of underlying carbonatized mafic volcanic rocks.
Upper Manitou Lake	Frenchman Island occurrence Swede Boys Island occurrence Gold Rock (or Haycock) occurrence	In quartz veins that occur in subvolcanic porphyry, intermediate pyroclastics, and mafic flows, that have been intruded by porphyry and felsite dikes.	Blackburn (1979a): Porphyries cited either (a) as heat source to concentrate gold, or (b) as carriers of residual gold at close of volcanic activity.
Manitou Island to Rector Lake	Gaffney prospect Bee-Hive prospect Royal Sovereign prospect Dryden-Red Lake prospect Swede Boy occurrence Reliance prospect Queen Alexandra occurrence Twentieth Century mine	In quartz veins in mafic metavolcanics, except for Gaffney (gold in sulphide zone) and Queen Alexandra (quartz vein probably in Carleton Lake Stock).	Blackburn (1976): Stratigraphic control in mafic volcanic sequence (Blanchard Lake Group) – possibly volcanogenic. Blackburn (1979a): Presence of porphyry and felsite dikes near most deposits suggests similar process to that at Upper Manitou Lake
Washeibemaga Lake	Pelham (or Forneri or Washeibemaga Lake) occurrence	In quartz veins and silicified zones at interior of thick pillowed basaltic flows.	Blackburn (1979b): Hydrothermal; emanation from Thundercloud porphyry along fault.
Mosher Bay	Giant prospect	Probably in quartz vein, in metasediments	Blackburn (1979b): Concentration by Taylor Lake Stock as heat source.
Mosher Bay	Big Dick occurrence	In quartz vein in mafic volcanic or intrusive rocks.	
Glass Bay	Glass Reef prospect	Probably in quartz veins in a sheared quartz porphyry.	Blackburn (1976): Stratigraphic control in mafic volcanic sequence (Wapageisi Lake Group).
Beck Lake	Watson and Wetelainen occurrence	Probably in quartz veins	

the theory that mafic volcanic rocks are source for gold (cf. Mackasey *et al.* 1974; Trowell *et al.* 1980; Pyke 1975). However, in considering the correlation of the Blanchard Lake group with the Wapageisi Lake group, as presented in this report, it might then appear inconsistent that only two or three deposits are known in the latter group.

Within the map-area 19 deposits are indicated northwest of the Manitou Straits Fault, compared with only five southeast of it. Of these latter five, one (Pelham occurrence near Washeibemaga Lake) is far removed. These observations suggest that a structural control might be considered. As noted in previous reports (Blackburn 1976, 1979a), minor structures such as foliation, lineation, etc. are far better developed northwest of the fault, and this is correlatable with the tight fold structure (Manitou Anticline). It is possible that the deformation opened up dilatent fractures favourable to emplacement of quartz veins. Although not the primary reason for location of these gold deposits, such fractures may have better enabled circulation and/or deposition of ore-bearing fluids.

All the above discussion suggests to the present author that the source of gold in the area was volcanic and subvolcanic rocks, and that concentration into presently known deposits was accomplished both by thermal or hydrothermal effects of subvolcanic or epizonal felsic intrusions and by opening up of dilatent zones during tectonism. The volcanic and subvolcanic sources were probably both mafic flows and sills, and felsic stock-like and sill-like bodies. Process of final emplacement is thus viewed as epigenetic, and may be compared with those discussed by Boyle (1979) under the heading of metamorphic secretion theories (p.395-398). There seems to be little evidence for syngenetic emplacement, although the hypothesis presented by Trowell *et al.* (1980) may have allowed for the primary concentration of gold at Gold Rock.

Base Metal Sulphide Deposits

Table 8 groups base metal sulphides known from drilling and surface prospecting in the Manitou Lakes map-area, and summarizes their association with geologic features.
The deposits found in three diamond drill holes between Merrill Lake and Stony Island, and the one occurrence found in a diamond drill hole on Manitou Straits, may be associated. The former lie within the Stony Island basalt formation, and the latter close to the Beaverhead Island formation. Airborne and ground electromagnetic surveys have located conductors that parallel stratigraphy in both of these formations, in the former from Merrill Lake to Stony Island, and in the latter discontinuously from Lower Manitou Lake to the northeast end of Manitou Straits. The author has suggested (Blackburn 1979a) a stratigraphic control for the occurrences between Merrill Lake and Stony Island, and it is suggested here that the Beaverhead Island formation may also be a stratigraphically favourable unit for such occurrences. From the stratigraphic analysis presented in this report it can be seen that both of these formations, on opposite limbs of the Manitou Anticline, are at the base of the Pincher Lake group, so that the base of this group in general may be considered a favourable location for base metal sulphide deposits.

Conductors located within the Glass Bay basalt formation have to date yielded little in the way of base metal sulphides. However, both Kerr Addison Mines Limited, and Canadian Nickel Company Limited have located conductors here, and a trace of chalcopyrite was reported from one of the former company's diamond drill holes.

At the base of the Wapageisi Lake group, minor chalcopyrite and sphalerite have been recorded in drill holes, and also by trenching and stripping in formation 1 of the Starshine Lake sub-group. As discussed by the author (Blackburn 1979b), other conductors may be located near the base of this thick mafic volcanic sequence, and formation 1 in particular may be a favourable stratigraphic level.

All of the above occurrences are considered here to be stratigraphically controlled. It is of note that all are contained within predominantly mafic metavolcanic, rather than felsic metavolcanic, sequences.

TABLE 8. EXPLORATION, MINERALIZATION AND ASSOCIATION OF BASE METAL SULPHIDE DEPOSITS IN THE MANITOU LAKES AREA.

Location	Exploration by[1]	Date	Method[1]	Mineralization and Association	Reported in
Merrill Lake to Stony Island	Kerr Addison Mines Ltd. Freeport Canadian Explor. Ltd.	1967 1970	Ground EM, drilling. Airborne EM and mag.	Minor chalcopyrite and sphalerite found with pyrite-pyrrhotite in silicified and altered zones in mafic metavolcanics. Stratigraphic control and volcanogenic origin suggested (Blackburn 1979a).	Blackburn (1979a).
Manitou Straits	Kerr Addison Mines Ltd. Freeport Canadian Explor. Ltd.	1967 1970	Ground EM, drilling. Airborne EM and mag.	Weak chalcopyrite in pyrite and graphitic shear zones in mafic metavolcanics.	Blackburn (1976): drill logs that describe mineralization were not included in 1976 report.
Glass Bay	Kerr Addison Mines Ltd. Canadian Nickel Co. Ltd.	1967 1971	Ground EM, drilling. Drilling.	Trace of chalcopyrite associated with pyrite-pyrrhotite and graphitic zones in mafic metavolcanics	Blackburn (1976): drill logs that describe mineralization were not included in 1976 report.
Boyer Lake	Massval Mines Ltd. Lynx Canada Explor. Ltd., Dejour Mines Ltd. Newmont Mining Corp.	1959 1970 1974	Drilling. Airborne and ground EM & mag, prospecting. Airborne and ground EM & mag, drilling.	Minor chalcopyrite found with pyrite along contact between gabbro sill and mafic metavolcanics and in graphitic zones in the volcanic rocks.	Blackburn (1979b).
Secret Lake	Canadian Nickel Co. Ltd. Lynx Canada Explor. Ltd., Dejour Mines Ltd.	1970 1970	Drilling. Airborne and ground EM & mag, prospecting.[2]	Chalcopyrite and sphalerite (minor) associated with pyrite and pyrrhotite in metagabbro and graphitic schist.	Blackburn (1979b).

[1]According to information in Assessment Files Research Office, Ontario Geological Survey, Toronto.
[2]According to a report supplied to the author by D.R. Derry, Consulting Geologist, Toronto.

A further group of occurrences are those at Boyer Lake, that are associated with a differentiated gabbroic sill (McMaster 1975), as reported by Blackburn (1979b).

It is of note, therefore, that all known base metal sulphide occurrences in the present area are associated with mafic rather than felsic rocks, be they volcanic or intrusive. Comparison may be drawn between this area and the Atikwa Lake area (Davies 1973), where base metal sulphides are similarly in mafic rather than felsic phases. However, in this latter area felsic phases are rare, in contrast to Manitou Lakes.

Molybdenum Deposits

Molybdenum is known at two occurrences in the map-area, one near Kaminni Lake (Blackburn 1976), and the other near Navimar Lake (Blackburn 1979a). In both places molybdenite occurs in quartz and quartzose pegmatite veins, and is confined to a small area. The veins intrude granitic rocks of the Atikwa Batholith near Kaminni Lake, and dioritic rocks at the margin of the batholith near Navimar Lake. No molybdenum has been found in the country rocks (Blackburn 1976, p.74; 1979a, p.64).

As noted in Trowell *et al.* (1980), other molybdenite occurrences are known at the edge of the Doré Lake lobe of the Atikwa Batholith. They cannot be considered analogous to porphyry-type molybdenum deposits, and appear to represent residual hydrous phases of the batholith.

Iron Deposits

Iron deposits are of two types, sedimentary and magmatic.

Sedimentary iron deposits, in the form of thin magnetite ironstone beds interbedded with sandstone, are found within the Mosher Bay formation of the Manitou group, and in very minor amounts in the sandstone at the top of the Stormy Lake group. Because of their narrow width, short strike length, and interbedded clastic sediments, none of these iron deposits are of present economic interest.

The association of magnetite ironstone, or "oxide facies" in the terminology of Gross (1965), with turbidites and resedimented conglomerates, indicative of deep submarine environment, is not unusual. Teal (1979) has noted the association at Manitou Lakes. Shegelski (1978), in a study of iron formations in the Savant Lake – Sturgeon Lake region, showed that, contrary to conclusions of earlier workers (e.g. James 1954; Goodwin 1962, 1973), facies of iron formations cannot be used as paleo-depth indicators. Rather, the sedimentology of enclosing sediments indicates bathymetry.

Shegelski (1978) concluded that in the area he studied a volcanic fumarolic source for the iron is most likely, that it was put into solution in the hydrosphere, and that among other criteria, "the preferrential association of laminated chemical sediment with siltstone-rich turbidite members ... indicate pelagic accumulation of a chemical precipitate" (p.238). Given the general similarity between the Sturgeon Lake – Savant Lake region and the present area, the present author believes Shegelski's conclusions to apply to the present area also.

Iron deposits of apparent magmatic type occur in a gabbro sill at Mountdew Lake (Blackburn 1979b), a gabbro sill south of Sunshine Lake (Blackburn 1979b), and in diorite marginal to the Atikwa Batholith (Blackburn 1979a).

REFERENCES

ACSN
1961: Code of Stratigraphic Nomenclature; American Commission on Stratigraphic Nomenclature, published in Bull. American Assoc. Petroleum Geologists., Vol.45, p.645-665.
Bateman, P.C., Clark, L.D., Huber, N.K., Moore, J.G., and Rinehart, C.D.
1963: The Sierra Nevada Batholith: a Synthesis of Recent Work across the Central Part; United States Geological Survey, Prof. Paper 414-D, 46p.
Beard, R.C. and Garratt, G.L.
1976: Gold Deposits of the Kenora-Fort Frances Area, Districts of Kenora and Rainy River; Ontario Division Mines, Mineral Deposits Circular 16, 46p. Accompanied by Chart A, scale 1:253,440 (1 inch to 4 miles).
Bertholf, W.E.
1946: Graded Unconformity: Washeibemaga Lake Area, Ontario; Unpublished M.Sc. Thesis, Univ. of Chicago, 45p., map, scale 1:63,360.
Birk, W.D.
1978: The Nature and Timing of Granitoid Plutonism in the Wabigoon Volcanic-Plutonic Belt, Northwestern Ontario; Unpublished Ph.D. Thesis, McMaster University, 521p.
Birk, D. and McNutt, R.H.
1977: Rb/Sr Isochrons for Archean Granitoid Plutons within the Wabigoon Greenstone Belt, Northwestern Ontario: a Preliminary Evaluation; p.161-167 in Report of Activities, Part A, Geological Survey Canada, Paper 77-1A.
Blackburn, C.E.
1976: Geology of the Lower Manitou-Uphill Lakes Area, District of Kenora; Ontario Division Mines, Geoscience Report 142, 81p. Accompanied by Map 2320, scale 1:31,680.
1979a: Geology of the Upper Manitou Lake Area, District of Kenora; Ontario Geological Survey Report 189, 74p. Accompanied by Map 2409, scale 1:31,680.
1979b: Geology of the Boyer Lake-Meggisi Lake Area, District of Kenora; Ontario Geological Survey, Open File Report 5263, 194p. Accompanied by 2 maps, scale 1:15,840. (Published in final form in 1981 as Ontario Geological Survey Report 202.)
1979c: Kawashegamuk Lake Area, District of Kenora; p.35-37 in Summary of Field Work, 1979, by the Ontario Geological Survey, edited V.G. Milne, O.L. White, R.B. Barlow and C.R. Kustra, Ontario Geological Survey, Miscellaneous Paper 90, 245p.
Blackburn, C.E., Beard, R.C. and Rivett, Scott
1981: Kenora-Fort Frances Sheet, Kenora and Rainy River Districts; Ontario Geological Survey, Compilation Series, Map P.2443, scale 1:253,440.
Boyle, R.W.
1979: The Geochemistry of Gold and its Deposits (together with a chapter on geochemical prospecting for the element); Geological Survey Canada, Bulletin 280, 584 p.
Bruce, E.L.
1925: Gold deposits of Kenora and Rainy River District; Ontario Department Mines, Annual Report for 1925, Vol.34, pt.6, p.1-42.
Coleman, A.P.
1895: Gold in Ontario: its Associated Rocks and Minerals; Ontario Bureau Mines, Annual Report for 1894, Vol.4, p.35-100.
1897: Third Report on the West Ontario Gold Region; Ontario Bureau Mines, Annual Report for 1896, Vol.6, p. 71-124.
Davies, J.C.
1973: Geology of the Atikwa Lake Area, District of Kenora; Ontario Division Mines, Geological Report 111, 57p. Accompanied by Map 2273, scale 1:31,680.
Davies, J.C. and Watowich, S.N.
1958: Geology of the Populus Lake Area; Ontario Department of Mines, Annual Report for 1956, Vol.65, part 4, p.1-24, Accompanied by Map 1956-3, scale 1:31,680.
Davis, D.W., Blackburn, C.E., Trowell, N.F., and Edwards, G.R.
1980: Geochronology of the Savant-Crow Lakes Area, Western Wabigoon Subprovince, Northwestern Ontario; in Summary of Geochronologic Research, 1978-1979, edited by E.G. Pye, Ontario Geological Survey, Miscellaneous Paper 92.

Elbers, F.J.
1976: Calc-Alkaline Plutonism, Volcanism and Related Hydrothermal Mineralization in the Superior Province of Northeastern Manitoba; Canadian Institute Mining and Metallurgy Bulletin, Vol.69, No.771, p.83-95.
Ferguson, S.A., Groen, H.A., and Haynes, R.
1971: Gold Deposits of Ontario, Part 1, Districts of Algoma, Cochrane, Kenora, Rainy River, and Thunder Bay; Ontario Department Mines and Northern Affairs, Mineral Resources Circular 13, 315p.
Fisher, R.V.
1966: Rocks Composed of Volcanic Fragments and their Classification; Earth-Science Reviews, Vol.1, p.287-298.
Goodwin, A.M.
1962: Structure, Stratigraphy, and Origin of Iron Formations, Michipicoten Area, Algoma District, Ontario, Canada; Geological Society America Bulletin, Vol.73, p.561-586.
1965: Preliminary Report on Volcanism and Mineralization in the Lake of the Woods – Manitou Lake – Wabigoon Region of Northwestern Ontario; Ontario Department Mines, Prelim. Rept. 1965-2, 63p. Accompanied by Chart, scale 1:253,440.
1970: Archean Volcanic Studies in the Lake of the Woods – Manitou Lakes – Wabigoon Region of Western Ontario; Ontario Department Mines, Open File Report 5042.
1973: Archean Iron-Formations and Tectonic Basins of the Canadian Shield; Economic Geology, Vol.68, No.7, p.915-933.
Gross, G.A.
1965: Geology of Iron Deposits in Canada; Vol.1: General Geology and Evaluation of Iron Deposits; Geological Survey of Canada, Econ. Geol. Rept. No.22, 181p.
Heimlich, R.A.
1971: Greenstone Assimilation by Tonalite Magma, Atikwa Lake, Ontario; Geological Magazine, Vol.108, No.1, p.1-80.
Irvine, T.N. and Baragar, W.R.A.
1971: A Guide to the Chemical Classification of the Common Volcanic Rocks; Canadian Journal of Earth Sciences, Vol.8, p.523-548.
James, H.L.
1954: Sedimentary Facies of Iron-Formation; Economic Geology, Vol.49, p.235-293.
Jensen, L.S.
1976: A New Cation Plot for Classifying Subalkalic Volcanic Rocks; Ontario Division Mines, Miscellaneous Paper 66, 22p.
Longstaffe, F.J.
1977: The Oxygen Isotope and Elemental Geochemistry of Archean Rocks from Northern Ontario; Unpublished Ph.D. Thesis, McMaster University, Hamilton, 590p.
Mackasey, W.O., Blackburn, C.E., and Trowell, N.F.
1974: A Regional Approach to the Wabigoon-Quetico Belts, and Its Bearing on Exploration in Northwestern Ontario; Ontario Division Mines, Miscellaneous Paper 58, 30p.
McInnes, W.
1895: Report of Field Work; in Summary Report, Geological Survey Canada, Vol.8, p.45-49.
1896: Report of Field Work; in Summary Report, Geological Survey Canada, Vol.9, p.34- 43.
1897: Report of Field Work; in Summary Report, Geological Survey Canada, Vol.10, p.38-43.
1902: Manitou Lake Sheet; Geological Survey Canada, Map No.720, scale 1 inch to 4 miles.
McMaster, G.E.
1975: Petrography and Geochemistry of the Gabbro Lake Sill, Superior Province, Northwest Ontario; Unpublished B.Sc. Thesis, McMaster University, 132p.
1978: Archean Volcanism and Geochemistry, Washeibemaga-Thundercloud Lakes Area, Wabigoon Subprovince, Superior Province, Northwest Ontario; Unpublished M.Sc. Thesis, McMaster University, 222p.
Moorhouse, W.W.
1939: Geology of the Eagle Lake Area; Ontario Department Mines, Vol.48, pt.4, p.1-31, Acc. by Map 48d, scale 1:63,360.
ODM-GSC
1961: Upper Manitou Lake, Kenora District, Ontario; Map 1153G, scale 1 inch to 1 mile. Survey 1961.
Parsons, A.L.
1911: Gold Fields of Lake of the Woods, Manitou and Dryden; Ontario Bureau Mines, Vol.20, pt.1, p.158-198.

1912: Gold Fields of Lake of the Woods, Manitou and Dryden; Ontario Bureau Mines, Vol.21, pt.1, p.169-204.
Pettijohn, F.J.
1937: Early Precambrian Geology and Correlational Problems of the Northern Subprovince of the Lake Superior Region; Bulletin Geological Society America, Vol.48, p.153-202.
1957: Sedimentary Rocks; Harper and Brothers, 718p.
Pichette, R.J.
1976: Petrology and Geochemistry of the Taylor Lake Stock, Superior Province, Northwest Ontario; Unpublished B.Sc. Thesis, McMaster University, 99p.
Poldervaart, A., and Parker, A.B.
1964: The Crystallization Index as a Parameter of Igneous Differentiation in Binary Variation Diagrams; American Journal Science, Vol.262, p.281-289.
Pyke, D.R.
1975: On the Relationship of Gold Mineralization and Ultramafic Volcanic Rocks in the Timmins Area; Ontario Division Mines, Miscellaneous Paper 62, 23p.
Riley, R.A., King, H.L. and Kustra, C.R.
1971: Mineral Exploration Targets in Northwestern Ontario; Ontario Department Mines and Northern Affairs, Miscellaneous Paper 47, 72p.
Sabag, S.F.
1979: The Geochemistry and Petrology of Granitoids at Meggisi Lake, N.W. Ontario; Unpublished M.Sc. Thesis, University of Toronto, 283 p.
Sage, R.P., Breaks, F.W., Stott, G.M., McWilliams, G.M., and Ali, A.
1974: Operation Ignace-Armstrong, Mine Centre - Entwine Lake Sheet; Districts of Kenora and Rainy River; Ontario Division Mines, Prelim. Map P.965, Geol. Ser. Scale 1:126,720.
Satterly, J.
1943: Geology of the Dryden-Wabigoon Area; Ontario Department Mines, Annual Report for 1941, Vol.50, part 2, p.1-67. Acc. by Map 50e, scale 1 inch to 1 mile.
Shegelski, R.J.
1978: Stratigraphy and Geochemistry of Archean Iron Formations in the Sturgeon Lake – Savant Lake Greenstone Terrain, Northwestern Ontario; Unpublished Ph.D. Thesis, University of Toronto, 251p.
Teal, P.R.
1979: Stratigraphy, Sedimentology, Volcanology and Development of the Archean Manitou Group, Northwestern Ontario, Canada; Unpublished Ph.D. Thesis, McMaster University, 310p.
Teal, P.R., and Walker, R.G.
1977: Stratigraphy and Sedimentology of the Archean Manitou Group, Northwestern Ontario; p.181-184 in Report of Activities, Part A, Geological Survey Canada, Paper 77-1A.
Thomson, J.E.
1934: Geology of the Manitou-Stormy Lakes Area; Ontario Department Mines, Annual Report for 1933, Vol.42, pt.4, p.1-40. Acc. by Map 42c, Scale 1:63,360.
1936: Gold Deposits of the Belt Extending from Manitou Lake to Lake of the Woods; Canadian Institute Mining and Metallurgy, Transactions, Vol.39, p.686- 701.
1942: Some Gold Deposits near Gold Rock, Upper Manitou Lake; Ontario Department Mines, Annual Report for 1938, Vol.47, pt.6, p.1-10. Accompanied by Map 47k, scale 1:4,800.
Trowell, N.F., Blackburn, C.E., Edwards, G., and Sutcliffe, R.H.
1977: Savant Lake-Crow Lake Special Project, Districts of Thunder Bay and Kenora; p.29-50 in Summary of Field Work, 1977, by the Geological Branch, edited by V.G. Milne, O.L. White, R.B. Barlow, and J.A. Robertson, Ontario Geological Survey, Miscellaneous Paper 75, 208 p.
Trowell, N.F., Blackburn, C.E., and Edwards, G.R.
1980: Preliminary Geological Synthesis of the Savant Lake-Crow Lake Metavolcanic-Metasedimentary Belt, Northwestern Ontario, and Its Bearing upon Mineral Exploration; Ontario Geological Survey, Miscellaneous Paper 89.
Turner, C.C., and Walker, R.G.
1973: Sedimentology, Stratigraphy, and Crustal Evolution of the Archean Greenstone Belt near Sioux Lookout, Ontario; Canadian Journal Earth Sciences, Vol.10, p.817-845.
Walker, R.G.
1976: Facies Models: 2. Turbidites and Associated Coarse Clastic Deposits; Geoscience Canada, Vol.3, p.25-36.
Wanless, R.K.
1970: Isotopic Age Map of Canada; Geological Survey Canada, Map 1256A, Compilation 1969.

INDEX

Alluvial fan environment ... 21
Amphibolite facies ... 45
Amygdaloidal
  Mafic metavolcanics ... 18, 26
Analyses
  Chemical
    Beck Lake section ... 28, 32, 34
    Blanchard Lake group ... 28-29, 33, 36
    Boyer Lake group ... 30
    Breccia ... 43
    Cane Lake formation ... 40
    Carleton Lake Stock ... 43, 44
    Doyle Lake section ... 28-29, 33, 36
    Gabbro Lake sill ... 30
    Granitic rocks ... 42-45
    Granodiorite ... 43
    Granodiorite, equigranular ... 43
    Jonas Lake section ... 29, 33, 37
    Meggisi Pluton ... 43, 44
    Metavolcanics ... 27-42
    Monzonite ... 43
    Pincher Lake group ... 29, 31, 33, 37
    Porphyry ... 43
    Quartz monzonite ... 43
    Quartz monzonite, seriate ... 43
    Scattergood Lake Stock ... 43, 44
    Stormy Lake group ... 40
    Starshine Lake section ... 28, 32, 35
    Sunshine Lake formation ... 38, 39
    Sunshine Lake subvolcanic intrusion ... 40, 43
    Syenodiorite ... 43
    Taylor Lake Stock ... 43, 44
    Thundercloud Lake porphyry ... 40, 43
    Upper Manitou Lake group ... 31
    Wapageisi Lake group . 28, 30, 32, 34, 35
Arenite ... 9-10
  see also: Sandstone
Argillite ... 21, 36
Atikwa Batholith ... 7, 12, 14, 42, 50, 55
Atikwa Lake ... 7

Basalt ... 25
  Alkalic, trachybasalt ... 49
  Pillowed to massive ... 6, 18, 19
    Porphyritic ... 18, 19
  Tholeiitic ... 49
Base Metal Sulphides ... 53-55
Beaverhead Island ... 10, 24, 46, 47
Beaverhead Island formation ... 17, 26, 53
Beck Lake ... 18, 19, 52
Beck Lake section, Wapageisi Lake group
  Chemical analyses ... 28, 32, 34
Bedding ... 47
Bee-Hive prospect ... 52
Bending Lake ... 2, 48
Benson Bay ... 25
Benson Bay sub-group ... 17, 26

Big Dick occurrence ... 52
Big Manitumeig Lake ... 14, 26
Big Master mine ... 1, 52
Blanchard Lake group ... 17, 23-24, 41, 49
  Chemical analyses ... 28-29, 33, 36
  Gold, in ... 51, 53
Boyer Lake ... 6, 48, 54, 55
Boyer Lake group ... 17, 23, 41, 46, 49, 50
  Chemical analyses ... 30
Breccia
  Mafic metavolcanic ... 19
Breccia, autoclastic
  Subalkalic felsic metavolcanic ... 8, 21, 41

Canadian Nickel Co. Ltd. ... 53, 54
Cane Lake ... 20, 21, 49, 50
Cane Lake formation ... 17, 20, 35, 37
  Chemical analyses ... 40
Carleton Lake Stock ... 13, 14, 45
  Chemical analyses ... 43, 44
Chemical analyses
see: Analyses, chemical
Chert ... 10, 19
Conglomerate ... 9, 21, 22

Dacitic rock ... 8, 41
Dark Horse conglomerate ... 16, 22
Detola prospect ... 52
Diabase ... 15
Diorite ... 7, 42, 55
  Equigranular ... 14
  Iron ... 55
  see also: Quartz Diorite; Syenodiorite
Dog Lake formation ... 25
Dore Lake lobe ... 55
Dorothy Lake ... 18
Doyle Bay ... 24
Doyle Lake section, Blanchard Lake group .. 41
  Chemical analyses ... 28-29, 33, 36
Doyle Lake Sill ... 24
Dryden-Red Lake prospect ... 52

Etta Lake formation ... 17, 19
Elora (or Jubilee) mine ... 1, 52

Fluvial environment
  Braided ... 21
Foliation ... 47, 48
Freeport Canadian Explor. Ltd. ... 54
Frenchman Island ... 11
Frenchman Island occurrence ... 52
Frenchman Island subvolcanic intrusion ... 24, 42, 49

Gabbro Lake Gabbro ... 23, 41, 48, 50
  Chemical analyses ... 30

Gabbroic rock 11, 19, 23, 27
Base metal sulphides, in 54, 55
Crystalline 6
Iron, in 55
Gaffney prospect 52
Garnet Bay 25
Garnet Bay formation 25
Geochronologic data 49-50
Giant prospect 52
Glass Bay 19, 52, 54
Glass Bay formation 17, 19, 53
Glass Reef prospect 52
Gold 1, 2, 50-53
Genesis 52
Gold Rock 1, 2, 51-53
Gold Rock (or Haycock) occurrence 52
Granitic rock 11-15, 49-50
Chemical analyses 42-45
Granodiorite 14, 42, 44
Chemical analyses 43
Equigranular 14
Chemical analyses 43
Micro 11, 24
Greenschist facies 45

Harper Lake 26
Holcroft Lake 10

Intrusive rocks
Felsic to intermediate 11-15
Hypabyssal 11
Batholithic 11-14
Plutons 14-15
Mafic 15
Mafic to ultramafic 10-11
Irene-Eltrut Lakes Batholithic Complex 14
Iron 55
Iron Formation
Gold, in 51
Ironstone
Magnetite 10, 46
Iron 55

Jasper 10
Johnar Lake 26
Johnar Lake formation 26
Jonas Lake section, Pincher Lake group 42
Chemical analyses 29, 33, 37
Jubilee mine
see: Elora mine

Kabagukski Lake 46
Kakagi Lake 3
Kamanatogama Lake 48
Kamanatogama Syncline 48, 50
Kaminni Lake 55
Kaminni Lake lobe 14, 25
Kekekwa Lake 26
Kennewapekko Creek 10
Kenny Lake Gabbro 23, 50
Kerr Addison Mines Ltd. 53, 54

Lacustrine facies 21
Lamprophyre 11, 21
Laurentian mine 1, 52
Lineation 47, 48
Lithologic units, table 4-5
Little Master prospect 52
Lynx Canada Explor. Ltd.,
Dejour Mines Ltd. 54
Magnetite
Ironstone 10
Manitou Anticline 5, 8, 26, 45, 47-48, 50
Manitou Group 15, 19-20
Chemical analyses .... 35-39 passim
Iron, in 55
Manitou "Series" 3, 9, 15, 20
Manitou-Stormy Lakes metavolcanic-metasedimentary belt 3
Petrochemistry 27-42
Stratigraphy 15-27
Manitou Straits Fault 3, 46, 50, 51, 53
Stratigraphy of metavolcanic-metasedimentary belt
Southeast of fault 16-23
Northwest of fault 23-27
Manitumeig Lake formation 26
Marius Lake 26
Massval Mines Ltd. 54
Meggisi Lake 14, 18
Meggisi Lake lobe 14
Meggisi Pluton 12, 42, 49
Chemical analyses 43
Meridian Bay 10, 19
Merrill Lake 25, 47, 53, 54
Metasediments 8-10, 46, 49
Chemical 10, 19
Clastic 9-10, 19, 21-22
Gold, in 51
Iron, in 55
Metavolcanics 5-8
Chemical analyses 27-42
Felsic, subalkalic 7-8
Gold, in 51
Intermediate, subalkalic 7, 24, 25, 26
Mafic 18-19, 21, 23, 24, 25, 26
Base metal sulphides, in 53, 53
Gold, in 51, 53
Mafic, alkalic 8
Mafic, subalkalic 6-7
Mitchell Lake Gabbro 11, 27
Molybdenum 55
Monzonite 15, 44-45
Chemical analyses 43
see also: Quartz Monzonite
Mosher Bay 10, 11, 46, 52
Mosher Bay formation 17, 21-22, 36, 46
Iron, in 55
Mosher Bay Gabbro 23, 50
Mosher Bay-Washeibemaga Lake Fault .. 46, 50
Mountdew Lake 55
Mountdew Lake Gabbro 23, 48, 50
Mudstone 10

Navimar Lake ... 55
Newmont Mining Corp. ... 54
Noonan Lake formation ... 26
Noxheiatik Lake ... 48

Paymaster prospect ... 52
Pelham (or Forneri or Washeibemaga Lake) occurrence ... 52, 53
Peridotitic rock ... 10-11
Pillows ... 6, 18, 46, 48
Pincher Lake group ... 17, 24-27, 42, 47, 49,
  Base metal sulphides ... 53
  Chemical analyses ... 29, 31, 33, 37
  Gold, in ... 51
Plagioclase-phyric flows ... 6, 25
Porphyries ... 49
  Chemical analyses ... 43
  Dacitic ... 41
  Quartz-feldspar ... 11, 42
  Rhyolitic ... 41
Pyroclastic rocks ... 26, 49
  Chemical analyses ... 40
  Subalkaline felsic metavolcanic ... 8, 21, 22
  Subalkaline intermediate metavolcanic ... 7, 19, 20, 24
Pyroxenitic rock ... 10

Quartz Diorite ... 7, 42
  Equigranular ... 14
  Micro ... 11, 24
Quartz Monzonite ... 44-45
  Chemical analyses ... 43
  Porphyritic ... 14
  Seriate ... 14, 42
    Chemical analyses ... 43
Quartz veins
  Gold ... 52, 53
  Molybdenum ... 55
Queen Alexandra occurrence ... 52

Rector Lake ... 51, 52
Reliance Bay ... 24
Reliance prospect ... 52
Rhyolitic rock ... 8, 41, 49
Royal Sovereign prospect ... 52
Rush Bay ... 21

Sagenak conglomerate ... 16, 22
Sandstone ... 9-10, 21, 22, 36
Scattergood Lake ... 18, 19, 50
Scattergood Lake Stock ... 13, 14, 45, 50
  Chemical analyses ... 43, 44
Secret Lake ... 18
Selby Lake prospect ... 52
Shaughnessy Bay ... 24
Sills, mafic ... 21, 49
  Chemical analyses ... 38, 39
Starshine Lake ... 18, 19
Starshine Lake section, Wapageisi Lake group ... 34
  Chemical analyses ... 28, 32, 35
Starshine Lake sub-group ... 17, 18-19
  Base metal sulphides, in ... 53
Stony Island ... 53, 54
Stony Island formation ... 25
  Base metal sulphides, in ... 53
Stormy Lake group ... 17, 22-23, 40-41
  Chemical analyses ... 40
  Iron, in ... 55
Submarine fan environment ... 49
Sunshine Lake ... 8, 10, 18, 20, 21, 49, 55
Sunshine Lake formation ... 17, 21, 36-37
  Chemical analyses ... 38, 39
Sunshine Lake Subvolcanic Intrusion (or Porphyry) ... 20, 37, 49
  Chemical analyses ... 40, 43
Swede Boy (or Swede Boys Island) occurrence ... 52
Syenodiorite ... 15, 44-45
  Chemical analyses ... 43
Syncline ... 3, 5

Taylor Lake ... 18
Taylor Lake Fault ... 50
Taylor Lake Stock ... 13, 14, 15, 44-45, 50
  Chemical analyses ... 43, 44
Thundercloud Lake ... 8, 18
Thundercloud Lake subvolcanic intrusion ... 22, 37, 49
  Chemical analyses ... 40, 43
Trafalgar Bay ... 1
Tuffaceous rock ... 25, 26
  Subalkalic felsic metavolcanic ... 8
Twentieth Century mine ... 1, 52

Uphill Lake ... 20, 21
Uphill Lake formation ... 17, 21, 35, 36
Upper Manitou Lake group ... 17, 24, 42, 47
  Chemical analyses ... 31
  Gold, in ... 51

Victory (or Upper Neepawa) occurrence ... 52
Volcanic Reef (or Vulcan) prospect ... 52
Volcanism
  Alkaline ... 49
  Calc-alkaline ... 49

Wacke ... 9, 10
  see also: Sandstone
Walmsley Lake ... 48
Wapageisi Lake group ... 16-19, 49
  Base metal sulphides, in ... 53
  Chemical analyses ... 27-35 passim
  Gold, in ... 53
Washeibemaga Lake 8, 10, 11, 16, 22, 49, 52, 53
Watson and Wetelainen occurrence ... 52
Watson's Falls ... 21
Watson's Narrows ... 26

Xenoliths ... 7, 26